THE **NATO** ENLARGEMENT DEBATE, 1990–1997

THE WASHINGTON PAPERS

. . . intended to meet the need for an authoritative, yet prompt, public appraisal of the major developments in world affairs.

President, CSIS: David M. Abshire

Series Editor: Walter Laqueur

Director of Studies: Erik R. Peterson

Director of Publications: James R. Dunton

Managing Editor: Donna R. Spitler

MANUSCRIPT SUBMISSION

The Washington Papers and Praeger Publishers welcome inquiries concerning manuscript submissions. Please include with your inquiry a curriculum vitae, synopsis, table of contents, and estimated manuscript length. Manuscript length must fall between 30,000 and 45,000 words. All submissions will be peer reviewed. Submissions to *The Washington Papers* should be sent to *The Washington Papers*; The Center for Strategic and International Studies; 1800 K Street NW; Suite 400; Washington, DC 20006. Book proposals should be sent to Praeger Publishers; 88 Post Road West; P.O. Box 5007; Westport, CT 06881-5007.

THE WASHINGTON PAPERS/174

THE **NATO** ENLARGEMENT DEBATE, 1990–1997

Blessings of Liberty

Gerald B. Solomon

Published with the Center for Strategic
and International Studies, Washington, D.C.

 PRAEGER

Westport, Connecticut
London

Library of Congress Cataloging-in-Publication Data

Solomon, Gerald B. H., 1930–
      The NATO enlargement debate, 1990–1997 : blessings of liberty
/ Gerald B. Solomon.
            p.   cm. — (The Washington papers : 174)
      ISBN 0-275-96289-X (cloth). — ISBN 0-275-96290-3 (pbk.)
      1. North Atlantic Treaty Organization—Europe, Eastern.
I. Title.   II. Series.
UA646.3.S654   1998
355′.031091821—dc21   97-50393

The Washington Papers are written under the auspices of the Center for Strategic and International Studies (CSIS) and published with CSIS by Praeger Publishers. CSIS, as a public policy research institution, does not take specific policy positions. Accordingly, all views, positions, and conclusions expressed in the volumes of this series should be understood to be solely those of the authors.

British Library Cataloging in Publication data is available.

Library of Congress Catalog Card Number: 97-50393
ISBN: 0-275-96289-X (cloth)
        0-275-96290-3 (paper)

First published in 1998

Praeger Publishers, 88 Post Road West, Westport, CT 06881
An imprint of Greenwood Publishing Group, Inc.

Printed in the United States of America

∞™

The paper used in this book complies with the Permanent Paper Standard issued by the National Information Standards Organization (Z39.48-1984).

10   9   8   7   6   5   4   3   2   1

Contents

List of Abbreviations   ix

Introduction   1

1. **Falling Satellites**   6

The Hand of Friendship   7
The "Coup" against Gorbachev   12
The North Atlantic Cooperation Council   13
An Arms-Length Friendship?   15
Peacekeeping   17

2. **Beyond Cooperation**   19

Associate Allies?   19
Encounter in Warsaw and the Secret Letter   23
A Concrete Perspective   24

3. **Partnership for Peace**   26

A Fresh Breeze from Evere   26
The Allies   30
Travemünde   32

4. **The Brussels Summit**   37

Presentation   43
Aftermath   46

5.	**A Special Partnership**	**53**
	Main Parameters	56
	Consultations ''as Appropriate''	58
6.	**How and Why**	**62**
	Reassurance	62
	Puzzling Evidence	65
	A Process of Examination	70
7.	**From Noordwijk to Brussels**	**74**
	Cold Peace?	79
	The Enlargement Study	85
8.	**Intensified Dialogue**	**88**
	Warning Signs	92
	The NATO Enlargement Facilitation Act	99
9.	**Endgame Afoot**	**102**
	One or More	102
	Enhanced Partnership	106
	The NATO-Russia Act	111
	Ukraine	120
	Affordable Stability?	121
	Closing Arguments	123
10.	**From Paris to Madrid: The Defining Moment**	**129**
	Conclusion and Recommendations	**139**
	Appendixes	
	A. Madrid Declaration on Euro-Atlantic Security and Cooperation, July 8, 1997	143
	B. Chronology of Principal Events	148
	C. Fundamental Principles of NATO Enlargement	160
	D. Signatories to the Partnership for Peace	161

   E.   Benefits of NATO Enlargement   162
   F.   Focus of Pre-Accession Military Work   163

Notes   **164**

Selected Bibliography   **179**

Index   **183**

About the Author   **191**

List of Abbreviations

ACE	Allied Command Europe
APC	Atlantic Partnership Council
ARRC	ACE Rapid Reaction Force
CBO	U.S. Congressional Budget Office
CFE	Treaty on Conventional armed Forces in Europe
CINCEUR	U.S. Commander-in-Chief, Europe
CIS	Commonwealth of Independent States
CJTF	Combined Joint Task Force
CNAD	Conference of National Armaments Directors
CSCE	Conference on Security and Cooperation in Europe (1975-1994)
DOD	U.S. Department of Defense
DPC	Defense Planning Committee
EAPC	Euro-Atlantic Partnership Council
EC	European Communities
EU	European Union
FRG	Federal Republic of Germany
FYROM	Former Yugoslav Republic of Macedonia
GDR	German Democratic Republic
G-7	Group of Seven (industrialized nations)
IFF	Identification Friend or Foe
IFOR	Implementation Force (Bosnia-Herzegovina)
IPP	Individual Partnership Program
NAA	North Atlantic Assembly
NAC	North Atlantic Council
NACC	North Atlantic Cooperation Council
NATO	North Atlantic Treaty Organization
OSCE	Organization for Security and Cooperation in Europe (1995-present, formerly CSCE)
PARP	Planning and Review Process

PCC	Partnership Coordination Cell
PFP	Partnership for Peace
PMSC	Political-Military Steering Committee
SACEUR	Supreme Allied Commander, Europe
SACLANT	Supreme Allied Commander, Atlantic
SFOR	Stabilization Force (Bosnia-Herzegovina)
SHAPE	Supreme Headquarters Allied Powers Europe
SLG	Senior Level Group
STANAG	Standardization Agreement
UNSC	United Nations Security Council
USSR	Union of Soviet Socialist Republics
WEU	Western European Union

THE **NATO** ENLARGEMENT DEBATE, 1990–1997

Introduction

*The Parties may, by unanimous agreement, invite any other Euro-
pean state in a position to further the principles of this Treaty and to
contribute to the security of the North Atlantic area to accede to this
Treaty.*

Article 10, North Atlantic Treaty
April 4, 1949

On July 8–9, 1997, NATO broke silence on a vital issue of post-
Cold War European security. At the NATO summit in Madrid,
which I attended as a member of a congressional delegation, alliance leaders decided that the time had come to offer to the new
democracies of Central Europe the very same democratic and
integrative security and political benefits enjoyed by the 16 relatively prosperous, and now threat-free, NATO allies. The Czech
Republic, Hungary, and Poland were invited to begin accession
negotiations with the aim of alliance membership by 1999, the
fiftieth anniversary of NATO. The alliance door would remain
open to other prospective members, while NATO would continue
to strive to forge cooperative security relationships with all European and Eurasian states, including a strategic partnership with
Russia to address common security challenges and draw Russia
closer to the West. (See appendix A, this volume.)

This event marked a watershed in post-Cold War history.
Since the early 1990s, many of the former Warsaw Pact member
states, emerging from more than four decades as satellites of a
totalitarian USSR, viewed membership in NATO as well as the

European Union as essential to securing their reintegration into the "West." They regarded membership as a hedge against the kind of Great Power rivalry and adventurism that had so scarred their history before and after World War II and robbed them of the Marshall Plan.

NATO had enlarged three times since its founding in 1949. The new European democracies believed they too were entitled to become part of an alliance dedicated to safeguarding "the freedom, common heritage and civilization of their peoples, founded on the principles of democracy, individual, and the rule of law," as the preamble to the 1949 North Atlantic (or Washington) Treaty solemnly declares. Membership in the most successful alliance in history, whose purpose did not fade even when the immediate post-World War II military task of containing Soviet adventurism had passed, could do for the new democracies of Central Europe what NATO had done for Western Europe: encourage irreversible democratic reform; foster patterns of cooperation and good neighbor relations; provide a secure climate for growth, trade, and investment; forge transparent and integrated defenses that would avoid potentially disastrous renationalization of defense, including nuclear, policy; curb possible tendencies toward violent disintegration along ethnic or territorial lines; strengthen NATO's ability to support security in Europe by collective defense against an attack upon alliance territory or crisis management outside its territory in defense of wider European security interests; widen the transatlantic partnership that history taught was indispensable to security and stability; and deter temptations by any state to impose hegemony.

NATO membership would, in short, contribute significantly to anchoring reform and returning these nations to a Europe based on shared, democratic values. It could eliminate a perceived perilous political and security "grey zone" in the same region that sparked two world wars in the twentieth century, thus supporting the vital interests of all European states and directly advancing U.S. national security interests.

The outcome of this grand experiment is unforeseeable, for it is largely an act of faith based on an intuition of a better future. Certainly, the road to Madrid had proved extraordinarily contentious, and the summit has crossed only the first, however important, threshold in defining the new NATO. Many questions persist: Despite the effort to engage Russia as a partner, could enlargement provoke a Russia in transition and thus leave all states in a worse position than before? Although not all of the 12 nations currently seeking NATO membership can join at the same time, is there a risk of drawing "dividing lines" and undercutting the course of reform in countries not admitted for the foreseeable future, or even implying a wholly unacceptable NATO-Russia understanding about comparative "predominance" in Central and Eastern Europe—thus turning back the clock to Yalta? Are future members truly prepared to accept the responsibilities of NATO membership, and are present allies ready to extend security pledges and expend resources to help integrate new members? How will a larger alliance contribute to ensuring an equitable sharing of responsibilities in the pursuit of all of NATO's missions and avoid the "free rider" phenomenon, and how can we determine "fair shares" in the absence of the "glue" of a Cold War clear and present danger? What kind of organization is being enlarged—a collective defense organization for only some nations? Or is the seed being sown for a comprehensive new collective security *and* defense system that includes all of Europe and its periphery far beyond the "North Atlantic area"? And most fundamentally, what beyond the immediate enlargement will be required for NATO to realize its ultimate political purpose, defined 30 years ago, of achieving "a just and lasting peaceful order in Europe accompanied by appropriate security guarantees"?[1]

All of these and many other questions have increasingly come to preoccupy NATO governments, parliaments, and policy elites as the enlargement process has unfolded. Countless editorials have addressed the "if," "how," "why," "when," and "who" dimensions of a process always held out under Article 10 of the North Atlantic Treaty—NATO's founding document, which is also known as the Washington Treaty. These issues will continue to generate debate despite the Madrid summit decisions and will invariably influence legislators in discharging their historic responsibility to provide advice and consent to ratification before April 1999 of the protocols of accession. This short history is intended to help place these issues in perspective, answer the skeptics of enlargement, and provide the missing historical compendium on the most profound geopolitical challenge of European security on the cusp of the twenty-first century.

Chapter 1 reviews NATO's initial response, from 1989 to

1990, to the collapse of the Warsaw Pact. In 1990 NATO extended a "hand of friendship" to allow for dialogue and cooperation with all the Warsaw Pact states. To those seeking the closest possible relations with the alliance on their way toward full membership—namely, the then Czech and Slovak Republic, Hungary, and Poland—the undifferentiated nature of this outreach failed to meet their high expectations. However, these nations took advantage of limited opportunities to become more familiar with the ways and means of the alliance. It was in part from this wide-scale cooperation with partner nations that NATO, seeking to ensure its relevance, began preparing for peace support operations outside its treaty area. The ability to act outside the treaty area was never ruled out by the Washington Treaty, but had always proved highly contentious, as the attempt to achieve consensus on the alliance role in the former Yugoslavia demonstrates. With the disappearance of the risk of full-scale attack on NATO's fronts, however, the current most likely risks are "spillover" from conflicts on its periphery. Because early action to contain and defuse these conflicts is in NATO's direct interests, outreach thus helped to spawn NATO's own transformation.[2]

Chapter 2 explores the early moves from outreach toward enlargement. Although the alliance took no view on this issue, by the end of 1993 a "perspective" was clearly opening.

Chapters 3, 4, and 5 review how NATO sought to combine the two strands of prospective enlargement while engaging nations not seeking NATO membership, especially Russia, to prepare for coalition operations and spread democratic security values such as civilian control of the military with the Partnership for Peace (PFP) in 1994. The PFP signaled that the alliance would "expect and welcome" new membership and introduced much more concrete military cooperation measures that could help prepare nations to join. Yet it engendered a sharp debate about whether NATO was avoiding the enlargement issue either to avoid change or appease Russia—even though as a defensive alliance NATO had no reason to defer to any other state. At the same time, the alliance agreed to a Russian search for a "special partnership" with NATO, even while Russia continued in its diplomacy to "transform" NATO away from a defensive alliance toward a component of an undefined pan-European security system and certainly to block its enlargement.

Chapters 6 through 10 trace how the PFP, while maintaining

its importance, eventually progressed toward the decisions taken at the Madrid summit. The chronology (appendix B, this volume) then traces these principal events for ease of reference.

In preparing this volume, extensive reliance was placed on the valuable North Atlantic Assembly (NAA) Reports, 1990–1997, by Bruce George MP (United Kingdom), Karsten Voigt (Germany), Jan Petersen (Norway), Guido Gerosa (Italy), Maurice Blin (France), Porter Goss (United States), Annette Just (Denmark), Longin Pastusiak (Poland), Petre Roman (Romania), Vyacheslav Nikonov (Russia), and Maarten van Traa (the Netherlands). John Borawski, director of the NAA Political Committee, contributed working papers for each chapter.

Many NATO and national diplomats, officers, and officials and congressional staff members provided accuracy checks and diverse perspectives. Special thanks go to Viorel Ardeleanu, Eitvydas Bajarunas, Karoly Banai, Major Craig Bell, Ian Brzezinski, Jim Doran, Lt. Colonel Herbert Harzan, Anne-Else Højberg, Hana Hubackova, Clarence H. Juhl, Gabor Iklody, Ratislav Kacer, Sergei Karaganov, Masha Khemelevskaya, Rick Kirby, Toivo Klaar, Wojciech Lamentowicz, David Law, Marcel Leroy, Kirsten Madison, Colonel James Mault, Ken Myers, Stefan Müller, Jan Nowak, Peter Michael Nielsen, Hans Jochen Peters, Colonel Jack Petri, Timo Pesonen, General Jiri Pislak, Robert Pszczel, Eszter Sandorfi, Thomas Szayna, Sergei Rogov, Jamie Shea, Jeffrey Simon, Colonel James Slocombe, Ed Timberlake, Jerzy Wieclaw, Witold Waszczykowski, and Thomas-Durrell Young.

Appreciation is wholeheartedly extended to CSIS for having made this publication not only possible but timely and to CSIS counselor Zbigniew Brzezinski for his constant encouragement and vision.

Finally, special gratitude is owed to CSIS's James R. Dunton, director of publications, and Donna Spitler, managing editor of the Washington Papers, for their invaluable assistance in the preparation and production of this volume.

The views expressed are solely the author's.

Washington, D.C.
August 1997

1

Falling Satellites

It was a permanent Soviet article of faith to link the future of the Warsaw Pact with NATO's. Indeed, unlike the North Atlantic Treaty, the 1955 Warsaw Pact Charter justified itself by the existence of an opposing "bloc." The Warsaw Treaty cited a "remilitarized Western Germany" in NATO as having increased "the danger of another war" (when in fact the Western intent was just the opposite), such that "the peace loving states of Europe should take the necessary measures for safeguarding their security." Also, unlike the North Atlantic Treaty, the Warsaw Treaty defined the condition for its expiration—the coming into force of a "general European Treaty of collective security."

Identical Soviet propaganda came into play 34 years later as it became clear that Germany would unify in NATO. Initially, in March 1990, Soviet head of state Mikhail Gorbachev "absolutely excluded" German unification, proposing instead "associate membership" of a single Germany in *both* NATO and the Warsaw Pact and consultative committees between the two alliances. Soviet diplomacy even suggested, after the "2 + 4" negotiations began on May 5, 1990, that the then German Democratic Republic (GDR) should honor its Warsaw Pact commitments for 5 years. Soviet foreign minister Eduard Shevardnadze declared in 1989 that "one cannot seriously think that the status of the GDR will change radically whilst the status of the FRG [Federal Republic of Germany] will remain as it was." He proposed that "permanent structures" of perhaps "even supranational character" should be agreed in the then 35-state Conference on Security and Cooperation in Europe (CSCE) as part of "the deepening of the common European process and the decisive dismantling of the structure of military confrontation"—as if the imposed Warsaw Pact and the democratic NATO were like creations. In addition to building up the CSCE, the USSR also sought arms control measures applied to Germany and NATO's reorganization to decrease what Shevardnadze termed "its purely military emphasis."[1]

Of course, when the Warsaw Pact finally collapsed from internal pressures on July 1, 1991, NATO not only still stood but emerged reinvigorated as a bright pole of intense attraction to the former Soviet "allies," some of whom had considered but rejected the options of neutrality, strengthening the CSCE, or remaining only as political members of the Warsaw Pact. No all-European, inescapably amorphous collective security system was in sight, nor was there any Western constituency for it. On February 15, 1991, in Visegrad, Hungary, the leaders of the Czech and Slovak Federal Republic, Hungary, and Poland declared their agreement to work together to achieve their "total integration" into the European political, economic, security, and legislative order.

The only agreed resulting restriction on the territory of the former GDR was that "foreign forces and nuclear weapons and their carriers shall neither be stationed in this part of Germany nor transferred there," whereas Germany unilaterally offered to limit its forces to 370,000 troops; however, German forces in this area could be fully integrated with "military alliance structures," and "the right of the united Germany to belong to alliances with all the rights and duties arising from this is not affected by this treaty."[2] Once peacetime conditions were no longer present, therefore, NATO would reserve the right to respond to the threat or use of force against *any* part of Germany as the allies saw fit.

In May 1990, however, President Vaclav Havel of the Czech and Slovak Federal Republic had already predicted that NATO could "become the seed of a new European security system"—a potential that developed in the form of NATO's outreach to the East, the adoption of new "crisis management" missions outside the treaty area, and enlargement of the alliance itself.[3]

The Hand of Friendship

On May 6, 1991, NATO secretary general Manfred Wörner stated that the democracies of Central Europe "neither want to be neutral nor components of a buffer zone, and nor do we."[4] Yet, he also stated that NATO did not want a "shift of balance or an extension of its military borders to the east,"[5] arguing that "our security spills over and contributes to deterring the idea that use of force, directly or indirectly, might lead to results."[6] But *how* could the alliance avoid buffer zones with their inherent invitation for an outside party to fill a vacuum, and *why* was the NATO secretary general so quick to rule out a wider alliance in response? If NATO did not favor neutrality for these nations, what *did* it support?

The truth is that it was not just the Soviet Union and some other allies that sought to preserve the Warsaw Pact. (Warsaw initially sought to maintain a Russian military presence as a counterweight to Germany, and Sofia did not seek to leave the pact as it was experiencing a controversy with Istanbul over the Turkish minority in Bulgaria.) On October 18, 1989, Gorbachev informed then chairman of the GDR Socialist Unity Party Egon Krenz that British prime minister Margaret Thatcher, French president François Mitterrand, and Italian prime minister Giulio Andreotti all "started from the preservation of the realities of the postwar period, including the existence of two German states [and] did not want to disrupt NATO and the Warsaw Pact."[7]

Some days later, on November 1, 1989, shortly before the historic and irreversible breach of the Berlin Wall, Ambassador Henning Wegener, the assistant NATO secretary general for political affairs, observed at a North Atlantic Assembly (NAA) meeting in Bonn: "If the Warsaw Treaty Organization is to be reformed, it must presumably be more in the direction of increased political consultation rather than of tighter coordination. This would, in the Western view, result in the Warsaw Treaty becoming an instrument of stability among states which differ nowadays considerably in their internal development."[8]

Zbigniew Brzezinski, national security adviser to President Jimmy Carter, declared a month earlier: "The termination of the two Alliances would contribute to anarchy. Institutional cooperation between the two alliances would preserve geopolitical stability and preserve territorial stability in Europe."[9] Half a year later, Henry Kissinger argued that "the most realistic security system" would include Austrian-type neutrality for Czechoslovakia, Hungary, and Poland (the "Visegrad" nations).[10]

Was this the sum Western response to liberators such as Lech

Walesa and Vaclav Havel who helped "win" the Cold War on freedom's terms? Was this the apex of the long-standing Western policy of encouraging "differentiation" in Central Europe from Soviet policy? Did this obvious reluctance to see the Warsaw Pact go reflect concrete or abstract concerns about "stability," a need to calm the USSR, a perceived need to maintain a rationale for NATO, a bureaucratic obsession with the tidiness of Cold War-era arms control negotiations, or some combination? On December 12, 1989, U.S. secretary of state James Baker even seemed indifferent: "Whatever security relationships the governments of Eastern Europe choose, NATO will continue to provide Western governments the optimal instrument to coordinate their efforts at defense and arms control, and to build a durable European order of peace"—which, he said, included NATO's offering the new democracies "an appealing model"—not, however, a *home*.[11] (In fact, the United States was then encouraging the Visegrad countries to form their *own* security forum, even though all of these countries were experiencing precipitous declines in defense budgets; all deployed obsolete equipment and could not possibly provide for their own national security against a major threat without external assistance.) In any event, let the record show that when I questioned Professor Brzezinski on November 15, 1994, at the NAA Washington Annual Session, he explained his 1989 remarks as intended to reassure the USSR, albeit as "part and parcel of encouraging the progressive dissolution of Soviet power."

Within a year, however, NATO realized that it could not simply carry on as if the world were standing still. At their London summit on July 5–6, 1990, the allies declared: "The Atlantic Community must reach out to the countries which were our adversaries in the Cold War, and extend to them the hand of friendship." Proposed was a joint declaration proclaiming that the NATO and Warsaw Pact nations were no longer adversaries (adopted in Paris on November 19 that year), presentations by these governments to the North Atlantic Council (NAC), regular diplomatic liaison with NATO for one-way information—"to share with them our thinking and deliberations in this historic period of change," and military contacts.[12] NATO secretary general Wörner then visited Moscow in July 1990, the first ever official NATO visit to the USSR, *two years after* this pioneering step had first been taken by his NAA counterpart, Peter Corterier.

NATO's "hand of friendship" could be seen as a historically extraordinary gesture. Others considered NATO's response as disappointingly sterile. That more was being considered was revealed by U.S. secretary of defense Dick Cheney just before the London summit: "Whether or not some kind of *observer status* would be arranged for, or some kind of *associate status* would ultimately be appropriate is something the Alliance is going to have to address."[13] Certainly, those nations most eager to join NATO—the three Visegrad countries—sought, at a minimum, political participation in NATO and, as the Czech and Slovak Federal Republic proposed in September 1991, a treaty-based security arrangement with NATO. They also sought an explicit NATO declaration affirming the vital interest of the alliance in the stability of the new democracies going beyond the "negative" formula Secretary General Wörner offered in April 1991 in Prague: "we are in no way indifferent to their security."[14] And they resented always being briefed at NATO collectively, as if the Warsaw Pact still existed—or, in NATO eyes, *should* exist. Canadian secretary of state for external affairs Barbara McDougall recalled: "The initial step in London was itself controversial, and the process has been the subject of fierce debates within NATO ever since."[15]

A certain degree of institutional competition between NATO and European organizations ran concurrent with the NATO outreach debate. NATO's precursor, the Western European Union (WEU), had helped integrate the FRG into the alliance, thereafter assigning any military role to NATO, and by 1991 was seeking to carve out an operational role as both the defense arm of the EU and the European "pillar" of NATO. WEU secretary general Willem van Eekelen suggested on June 3, 1991, that NATO would likely "continue to be preoccupied by the situation in the Soviet Union." The WEU, however, could adopt a more flexible and selective approach toward "those countries that have made real progress toward a democratic and pluralistic form of government." He made a case, strongly supported by France, for the WEU's serving as the organization for crisis management and peacekeeping in the region, suggesting that this would be less provocative than action by NATO.[16] (As of this writing, however, the WEU has never undertaken tasks beyond monitoring embargoes and providing police forces.)

NATO did issue a somewhat more robust declaratory affirmation "after much debate" during the June 6–7, 1991, Copenhagen NAC meeting:

> Our own security is inseparably linked to that of all other states in Europe. The consolidation and preservation throughout the continent of democratic societies and their freedom from any form of coercion or intimidation are therefore of direct and material concern to us . . . we will neither seek unilateral advantage from the changed situation in Europe nor threaten the legitimate interests of any state.[17]

This language, which did not elaborate on what or whose "legitimate interests" NATO would not threaten, had been prompted by a mischievious Soviet Communist Party International Department report of January 1991, outlining a strategy to control Central Europe by precluding foreign forces and emulating the Finnish-Soviet relationship. In April Soviet deputy foreign minister Yuli Kvitsinsky declared:

> It goes without saying that there can be no return to the policy of domination in the Eastern European region for any nation. At the same time, the Soviet Union's legitimate interests in this region have historical and geopolitical roots and must be taken into account. Those who value peace and stability on our continent, and want them to be lasting and durable, understand very well that the Eastern European region under no circumstances should become a source of threat to the security of the USSR. It is equally clear that *there should be no foreign military bases or armed forces in this region.* [Emphasis added.][18]

The NATO Copenhagen language was intended to project "positive ambiguity" about whether NATO might take action if a NATO liaison partner were threatened—just as Yugoslavia under Tito was thought to have indirectly benefited from U.S. statements attaching importance to that republic's independence. To others, however, it was a troubling sign of an obvious hesitation on the part of alliance governments that could even be recklessly misleading in a crisis—for example, the empty "rollback" and "liberation" rhetoric of the first Eisenhower administration.

In any event, the Copenhagen language was not entirely new: the CSCE "Charter of Paris for a New Europe," adopted on

November 21, 1990, by all NATO member states, declared that "security is indivisible and the security of every participating State is inseparably linked to that of all the others." Yet, these words remained just that when the tragedy of the former Yugoslavia erupted in June 1991 and demonstrated a lack of resolve when the Europeans chose initially to address the crisis themselves—a lesson not lost upon the new democracies of Central and Eastern Europe.

In Copenhagen the NAC also agreed, as previously urged by the NAA in November 1990 and as elaborated by Secretary Baker and German foreign minister Hans-Dietrich Genscher on May 10, 1991, on a more active program of "partnership" for exchanging views and information on security policy, military strategy and doctrine, and related topics, participation in NATO training and educational facilities in Rome and Oberammergau and scientific and environmental programs, NATO information programs in these countries, and orientation visits at NATO headquarters.

The "Coup" against Gorbachev

A temporary lapse from formal NATO "nondifferentiation" toward the former Warsaw Pact states was at least suggested during the supposed "coup" attempt against Gorbachev in August 1991. The five non-Soviet representatives were personally briefed for the first time by the NATO secretary general, whereas the Soviet representative was briefed separately an hour later. The NAC declared that "as a token of our solidarity with the new democracies [versus *states*, the term used in the Copenhagen communiqué] of Central and Eastern Europe, we will develop ways of further strengthening our contribution toward the political and economic reform process within these countries.[19] This language was intended to signal the coup leaders that NATO would "keep [its] options open," as a participant recalled.

However, the events in Moscow also prompted a reappraisal of the pace of NATO outreach. According to McDougall:

> The coup in the USSR demonstrated to the proponents of NATO outreach that half-measures had not been enough, while the opponents saw in the coup justification of their caution. . . . The aftermath of the coup, however, changed

all the calculations. The other nations of Central and Eastern Europe became even more insistent in their demands of NATO, whereas the voices of opposition in the USSR were swept away.[20]

The North Atlantic Cooperation Council

On October 2, 1991, Secretary Baker and Foreign Minister Genscher again took the initiative by proposing "a more routine set of meetings . . . perhaps as a 'North Atlantic Cooperation Council.' " NATO should also offer "to commence planning with liaison countries for *joint action* on disaster relief and refugee programs, and pledging NATO's support for CSCE in dealing with these and other new security challenges in Europe [emphasis added]"—that is, moving from dialogue exclusively to joint activities and real-world operations.[21] In May 1989, the Warsaw Pact had actually appealed to the NATO states to establish a political dialogue, but NATO did not favor direct organizational contacts because of the coercive nature of the Warsaw Pact.

Also in October, the NAA urged NATO to recall that Article 10 of the Washington Treaty permitted new states to join NATO and to interpret the article as referring to nations in which functioning democratic institutions were safely entrenched and whose security posture was compatible with the alliance. It proved over two years ahead of the alliance and would maintain a momentum in the years ahead, urging that NATO's welcome to the new democracies of Central and Eastern Europe extend beyond the initial limited outreach activity alone.

The Baker-Genscher proposal for a North Atlantic Cooperation Council, the NACC, was endorsed at the NATO Rome summit in November 1991 and convened on December 20, 1991, in Brussels. The first meeting coincided with the dissolution of the USSR and the receipt of a message from Russian president Boris Yeltsin that raised "a question of Russia's membership in NATO . . . as a long-term political aim."[22] The first "Work Plan for Dialogue, Partnership and Cooperation" was adopted on March 26, 1992, drafted not with NACC partners but by NATO for NACC approval, (NACC was, after all, "NATO-centric," and this unequal partnership would come to irritate partners and not just Russia.) Intended to develop a more institutional relationship of "consultation and cooperation," it would focus on security and related issues, including defense planning, conceptual approaches to arms control, democratic concepts of civilian-military relations, civil-military coordination of air traffic control, military-civilian defense conversion, scientific and environmental issues, and dissemination of information about NATO in Central and Eastern Europe. Secretary Baker stated in Brussels at the December 1991 NAC meeting that the NACC would serve as the "primary consultation body" between NATO and liaison states on security and related issues, and that NACC "could play a role in controlling crises in Europe. It might, for example, serve as a forum for communicating NATO crisis responses to liaison states, as well as give liaison states access to NATO when necessary."[23] However, the Baker-Genscher proposal for joint operations was judged premature, sparking a protracted intra-alliance dispute about the NACC's appropriate role relative to, for example, the CSCE and WEU (see the next section).

On March 10, 1992, all of the former Soviet republics were admitted into the NACC just at the time the Visegrad countries were actively seeking a "privileged" status with NATO, thus bringing the total to 37 states. A key reason for this inclusiveness was to preserve the CFE Treaty via the informal NACC "High Level Working Group," which first met on January 10, 1992, and which proved successful in achieving that goal. In contrast, on June 19, 1992, the WEU chose to formalize its explicitly "differentiated" approach to the Central European and Baltic nations with a perspective of joining the EU to acquaint these countries "with the future security and defense policy of the European Union and find new opportunities to cooperate with the defense component of the European Union and with the European pillar of the Atlantic Alliance as these develop." Foreign and defense ministers would henceforth meet annually, and ambassadors more regularly, in a "Forum for Consultation."[24] At a November 1992 seminar in Heidelberg, a senior Hungarian diplomat stated that his nation "sometimes feels that we are caught in a crossfire between WEU and NATO," although, in his view, relations with the WEU "are much less substantial and have no content."

Excluded from WEU dialogue, Russia, however, had to be central to NATO if not WEU concerns. The draft Russian military doctrine did not appear to rule out a return to confrontation with the West. Published in the May 1992 issue of *Military Thought*

(*Voennaya Mysl*), the draft stated that Russia would regard as a direct military threat the advance of foreign troops into the territory of neighboring countries or the buildup of army and naval groupings near its frontier. Should that occur, Russia reserved the right to take "necessary measures to guarantee its own security." A Russian Foreign Ministry "special release" in late 1992 also referred to the "strategic" task of preventing "attempts to turn Eastern Europe into a kind of buffer that would isolate us from the West" and "the quite tangible attempts of Western powers to force Russia out of Eastern Europe."[25] Moreover, although this factor figured less in NATO European than in U.S. policy, Russia remained a major nuclear power, and reliance on nuclear weapons could well increase, given a combination of deteriorating relations with NATO and conditions in the Russian armed forces—dramatically evidenced in Chechnya two years later.

An Arms-Length Friendship?

By mid-1992 NACC was not even half a year old, yet the Americans were pressing for the NACC to do more. On June 17, 1992, President George Bush and President Boris Yeltsin jointly proposed in an "American-Russian Charter for Partnership and Friendship" that the NACC should be able to contribute to OSCE *peacekeeping*—something that would never have attracted consensus in the alliance at that time. At the June 4, 1992, NAC ministerial in Oslo, U.S. deputy secretary of state Lawrence Eagleburger urged that "we should be prepared to invite the NACC countries and neutral states to contribute to peacekeeping operations in partnership with NATO," begin "preparatory work to make our forces interoperable for peacekeeping missions," support "NACC observers" in Nagorno-Karabakh, and assist partners in civil emergency, civilian air traffic coordination (in the past civilian and military air traffic systems were not coordinated in Warsaw Pact countries), and defense conversion."[26]

Not all NATO nations, however, supported the NACC as taking on an operational role. France was not in favor, exhibiting preference for the WEU and CSCE, whereas in May 1992 the United Kingdom appeared to seesaw:

We see the NACC as a forum for frank discussions of security issues of concern to its members [although the very Partners who might be generating such concerns were always at the same table]. We do not believe that it should be a crisis management forum or have an operational role in defence [or] in possible CSCE peacekeeping operations, although it would clearly give NACC partners a forum in which to discuss their response to European security problems which might demand a response from the CSCE of a peacekeeping force.[27]

Nevertheless, U.S. permanent representative to NATO Reginald Bartholomew expressed undisguised concern on September 23, 1992:

We cannot allow cooperation within NACC to become a *largely symbolic effort,* rather than an integral, substantive new part of NATO itself . . . NACC's full potential . . . has yet to be realized. NACC has yet to grow from a forum for periodic joint consultation into a tool for delivering tangible assistance to countries whose old security policies have proven bankrupt. It has yet to develop from an arms-length friendship into a vehicle for joint action that the word 'partnership' should represent.[28]

He urged that NATO make its special resources and expertise available for crisis management, peacekeeping, retraining Central European military establishments, and defense conversion: "*Makework* will not sustain the transAtlantic partnership or build a common security framework encompassing the countries of Central and Eastern Europe [emphasis added]."[29]

As an illustration of what Ambassador Bartholomew had in mind, the first NACC "work plan" contained the "topic" of civilian-military air traffic management, but the corresponding "activity" was no more than a seminar. In contrast, in December 1992 the United States sold air safety equipment to the Hungarian air force, the first such U.S. transfer to a former Warsaw Pact nation. Likewise, partners frankly complained that although the NACC was a valuable forum, it suffered from numerous drawbacks: consultations took place "as a rule" only every two months, which gave the impression of presenting pre-cooked NATO decisions; excessive time was spent on the agenda; the Secretariat and some allies did not appear to express great interest in NACC development; and no resources were forthcoming to support wide-ranging "activity." The United States and Germany had moved with great speed to secure a united Germany in NATO, yet the NACC, diffused even further by the automatic admission of the former Soviet republics, seemed like a slow train with an unknown destination.

Peacekeeping

The peacekeeping "breakthrough" finally occurred at the end of 1992. The rationale for focusing on peacekeeping was to help develop democratically controlled armed forces, enhance mutual understanding of security policies, create the conditions for wider participation in "crisis management" in the European area, and build the same habits of cooperation that were the trademark of the alliance. The 1993 NACC Work Plan, adopted on December 18, 1992, called for consultations toward cooperation in preparing for such peacekeeping activities as joint planning, joint training, and "consideration of possible joint peacekeeping exercises"—a formula required to overcome the objections of only one of the NACC countries, France. The NACC "Ad Hoc Group on Cooperation in Peacekeeping" was established some eight months later on February 11, 1993. NATO also made a point, in a nod toward differentiation, that cooperation could take the form of activities agreed by all partners but carried out by only some of them.

This at least gave the NACC a potentially *operational*, real-world dimension, more than a "forum for dialogue." It had the virtue of choosing a benign form of military cooperation—peacekeeping—that could still build the interoperability a future NATO member state would require. Nevertheless, so tedious was the debate at the December 1992 NACC meeting that Secretary General Wörner, after having endured a long U.S.-French harangue about the desirable degree of consultation versus cooperation (concerning peacekeeping and civil emergency planning), exclaimed that he had arrived at "the end of my patience and intelligence."

But NATO decisions tended to follow, rather than shape, events. Whereas no NACC peacekeeping exercises were yet being "considered," NACC members Poland, the Czech Republic, Russia, and Ukraine were *already* cooperating with their NATO "partners" in Croatia and Bosnia in real-world activity in the UN Protection Force (UNPROFOR). A Russian-U.S. rescue exercise was held in the Laptev Sea in April 1993, and during the following June forces and observers from Poland, Estonia, Latvia, and Lithuania, as well as from Sweden and Finland, participated alongside naval forces of Denmark, Norway, the Netherlands, Germany, and the United States in the U.S.-sponsored invitational exercise *Baltops* in the Baltic Sea.

As 1992 drew to a close, Secretary General Wörner conceded that "at this very moment an extension of membership is not on the agenda,"[30] although his statement at least acknowledged an issue that he had not considered, as noted, a year and a half earlier. Still, countries like the Visegrad nations could see no alternative to NATO membership, and no such alternative could honestly be explained to them—certainly no declaration that their security was of "direct and material concern" to the alliance or that NATO's sheer existence contributed to regional stability (witness the unchecked anarchy in former Yugoslavia). As the director of the International Security Department of the Polish Ministry of National Defense, and one of its first civilian employees, Andrzej Karkoszka, put it in March 1993: "NACC remains what it is: a consultative body without any executive powers composed of deeply disparate states. . . . Because of this fact, Poland cannot consider NACC as a substitute for direct cooperation with NATO especially in the military area."[31]

2

Beyond Cooperation

Associate Allies?

Within months of the NACC inaugural, suggestions already were being made as to where the process could lead. "I think membership could come within the next decade, perhaps even sooner," U.S. NATO ambassador William Taft stated on July 13, 1992, "but the NACC is not in itself a step toward membership."[1] At the Oslo NAC meeting in June 1992, Secretary Eagleburger ventured:

> We ought to be considering the possibility of extending the Alliance. There [are] a number of conditions involved in that, not least of which is commitments to democracy . . . but the point is—and this is the U.S. that was speaking [at the meeting]—not that there was any common agreement within the Alliance today, but the U.S. was suggesting that at the proper time in the future it might be possible to extend the Alliance.[2]

In fact, by late 1990 NATO enlargement was already being considered in Washington as a follow-on to NACC.[3]

The first *entreé* in the open policy debate was provided in 1992 by Jeffrey Simon, senior fellow at the Institute for National Strategic Studies of the National Defense University in Washington, D.C., and a key collaborator with Department of Defense (DOD) officials in Partnership for Peace (PFP) conception implementation. The PFP was intended to go beyond N
alogue and cooperation" to offer a practical program o

planning and budgeting, joint exercises and operations with
NATO nations, and, for the first time, a link to eventual NATO
membership for states seeking it. Simon argued that unless liberal
democracies were established in the former Warsaw Pact zone,
then "much of the West's investment in the Cold War will be
squandered."[4] NATO would become irrelevant, the Central and
Eastern Europeans would be cast adrift in a sea of instability,
and the renationalization of defense might occur with historical
suspicions aroused toward Germany if the alliance failed to open
its door to the East.

Simon suggested associate NATO membership leading in
"five-to-ten years" to full membership—with Greece and Turkey
having enjoyed "observer" status in NATO prior to their full
admission in 1952. Associates would be able to observe the NAC
and Military Committee sessions and would receive practical advice on standardization and interoperability. Although Article 10
of the Washington Treaty simply describes possible new members as European democracies (overlooked for periods of Greek,
Portuguese, and Turkish history) able to contribute to the security
of the North Atlantic area, Simon urged that new members would
specifically have to commit to support freely elected governments, privatize the economy, observe ethnic minority rights,
adopt a military doctrine and resources consistent with "reasonable sufficiency," honor relevant arms control agreements, and
ensure civilian control of the military. (In the past, the military,
which was shrouded in policy and financial secrecy, owed its
loyalty to the Communist Party, and its military production decisions were decided not nationally but by the Tenth Directorate of
the Soviet General Staff.) If the West would or could not provide
eventual security guarantees to additional nations, Simon urged
support at least for "regional military and security cooperation"
among the Visegrad nations to develop closer links with NATO
and WEU, aiming at a regional system of collective security.

These ideas for differentiation, however, ran into deep objections within NATO governments as somehow at odds with the
objective of drawing Russia as rapidly as possible into the "Western mainstream," or even legalistic arguments that no provision
existed in the Washington Treaty for associate membership (although the lack of a treaty provision had not prevented the WEU
from creating such a status for European members of NATO who
were not members of the EU—Iceland, Norway, and Turkey). As

the United Kingdom foreign office noted in May 1992: "We do not favor the development of explicit differentiation."[5] A number of allies remained wary of damaging relations with some NACC partners, such as Russia, if NATO cooperated too closely and used the territory of its neighbors for training. This perpetual European caution and preoccupation with the former hegemon in the East tended to rob NACC of some credibility and encourage "safe" but value-limited projects.

Another approach was called the "Royal Road": membership in the EU should precede or coincide *grosso modo* with membership in NATO. According to its logic, EU membership would offer a certain stability and effectively certify the nation concerned as joining the West. At the November 1991 NATO Rome summit, however, it was confirmed that a European defense identity should serve both as an independent capability and as the European "Pillar" of NATO, such that states who joined the EU and its (theoretical) defense arm, the WEU, should be invited to become members of NATO. For example, Belgian foreign minister Willy Claes (who succeeded NATO secretary general Wörner on October 17, 1994) noted in October 1993 that from the EU point of view a new NATO member "must at least be a candidate for EU membership."[6] EU membership prior to NATO enlargement was also the "easiest way" to explain to Russia why others could join NATO before Russia, if ever, did.

But the "Royal Road" was not the only view. As German defense minister Volker Rühe suggested on September 15, 1993: "I consider it conceivable and also desirable that membership be more likely in security policy, that is, in NATO. . . . These countries probably need a good 10 years" for EU membership.[7] U.S. Republican senator Richard Lugar argued on June 24, 1993, that "the parallel [EU-WEU/NATO] accession option is the worst of all worlds; it is not only the back-door approach, but . . . reverses the priorities and gives the Europeans something of a veto over both NATO expansion and revitalization. NATO has to have its own criteria for expansion."[8] The "backdoor" approach referred to NATO's becoming involved in a situation with a nonmember state. If, say, Poland joined the WEU, then member states would be obliged to come to Poland's defense, which would invariably draw in NATO because of WEU dependence on such NATO (primarily U.S.) resources as power projection assets and intelligence.

Another particularly important contribution came from NATO spokesperson Jamie Shea in September 1993. In a personal capacity, Shea argued that NATO's two fundamental interests were not only to respond to negative security developments that the alliance could not deter, but also to anticipate and reduce those very same potential threats. Here NATO could either accept as new members "the countries of Central and Eastern Europe and take the risk of importing their instabilities . . . or it can shut them out, with the risks that these instabilities will spread over Alliance territory in any case."[9]

Consequently, Shea proposed a two-part protocol of accession to the North Atlantic Treaty for "essentially Poland, the Czech Republic and Hungary":

- "a commitment to observe certain political precepts; and
- 'Spanish-style' coordination agreements [Spain retains operational command over its forces and thus only 'coordinates' its military relationship with NATO] whereby they would accept the general principles of the alliance's Strategic Concept and clarify what type of military role they wish to play. Within five years, the Article 5 security guarantee would be formally extended, but in the interim these three countries would be able to participate fully in all NAC political consultations and participate in all out-of-area activities."[10]

At the same time, Shea saw a continued role for NACC provided it could recover from its "lack of focus regarding its fundamental purpose." He proposed an NACC-dedicated secretariat at NATO and a peacekeeping planning group at Supreme Headquarters Allied Powers Europe (SHAPE) that would run frequent command and field exercises to promote standardization and interoperability of equipment and plan for contingencies. Inevitably NACC would resemble "a type of security CSCE"—reassuring to those countries interested in joining NATO without marginalizing those who did not.[11]

Likewise, some months later former NATO secretary general Lord Carrington noted "NATO's obvious hesitation" in choosing between accepting countries "such as Poland and the Czech Republic" into NATO and offering cooperation to everyone with indirect security assurances to Central Europe. He preferred the first alternative.[12]

Encounter in Warsaw and the Secret Letter

Intensified discussion was prompted by a joint statement agreed to by Presidents Walesa and Yeltsin in Warsaw on August 25, 1993. The statement recorded, despite the reportedly animated objections of Yeltsin's aides, that the two presidents had discussed "the issue" of Poland's intention to accede to NATO, and that Poland's position was received "with understanding" by Yeltsin. It further recorded: "In perspective, a decision of this kind by sovereign Poland aiming at all-European integration is not contrary to the interests of other states including also Russia." In September, the German NATO ambassador stated that "the possibility of opening up the WEU and NATO . . . is no longer taboo"[13] (an interesting admission), and it was assumed that if there were any objections to enlarging NATO, the alliance nations could no longer point to Russia. As Longin Pastusiak, Foreign Affairs Committee deputy chairman of the Polish Sejm (Lower House of Parliament), recounts:

> The issue of Polish membership in NATO was raised by the Polish side during private talks between the two presidents, and later by the Polish side in the plenary meeting of the two delegations. President Yeltsin did not object to Polish membership in NATO, but . . . put membership 'in perspective.' The Polish government somewhat overinterpreted that declaration by saying that Russia did not object to Polish membership in NATO. When President Yeltsin left Warsaw, the Polish government appealed to NATO governments to open alliance doors to Poland's full membership.
>
> President Yeltsin immediately came under criticism in Russia from almost all political groups. Opposition to Yeltsin's statements was publicly expressed by Foreign Minister Andrei Kozyrev, by Defense Minister Pavel Grachev, by some other top-level officers, and by virtually all Russian political parties. The following reasons were expressed for excluding the Central and East European countries from NATO, as inclusion would: isolate Russia from the West; create a buffer zone between Russia and Western Europe; deprive Russia of markets for military equipment in the Central and Eastern European countries; threaten the democratization process in Russia; create new military threats to Russia; and cause Russia to lose its importance to the West.

The net result of this, particularly in Russian military circles, which President Yeltsin could not afford to antagonize, was the secret letter to the leaders of Great Britain, France, Germany, and the United States of mid-September 1993.[14]

In that letter, President Yeltsin argued that "the spirit of these stipulations," which in the 2 + 4 treaty on German unification precludes the stationing of foreign forces in the eastern federal *Länder* of Germany, "rules out any possibility of a NATO expansion eastwards." Instead, Russia and NATO should "officially offer [Central European states] security guarantees . . . enshrined in a political declaration or a *treaty on cooperation between the Russian Federation and NATO* [emphasis added]." He also urged "another course of action . . . leading to a truly pan-European security system (not, however, on the basis of blocs)," while at the same time declaring that "relations between our country and NATO should be several degrees warmer than the relations between the alliance and Eastern Europe," and that a NATO-Russia relationship "could progress at a much quicker pace." He also stated that, although such a relationship was "presently . . . purely theoretical," Russia could join NATO "from a long-term point of view."[15] The result was another confusing message: in the longer term, NATO should not exist, but could exist, and Russia might join; blocs would have no role, but Russia and NATO should offer "cross-guarantees."

A Concrete Perspective

Even if the Yeltsin letter had not been dispatched, there was no consensus that enlargement was ripe for the forthcoming NATO summit proposed in May 1993 by the United States. As a result of conversations with Central and Eastern European leaders during the opening of the Holocaust Museum in Washington on April 22, President Clinton had opened the policy path regarding NATO enlargement. By the fall, Secretary General Wörner, by this time personally and even fervently committed to steering through a "transformed" alliance, urged on September 10, 1993, that the summit provide a "concrete perspective":

> A major, and perhaps the primary, future mission of NATO will be to project stability to the East . . . we should now

consider further steps. NATO is not a closed shop. . . . In my view, *the time has come to open a more concrete perspective to those countries of Central and Eastern Europe which want to join NATO and which we may consider eligible for future membership.* . . . Even if there are no immediate plans to enlarge NATO, such a move would increase the stability of the whole of Europe and be in the interests of all nations, including Russia and Ukraine [emphasis added].[16]

Yet, on October 9, 1993, Danish foreign minister Niels Helweg Petersen informed the NAA Political Committee in Copenhagen that although "NATO governments have agreed that the time is ripe for serious analysis" of admitting new alliance members, "we have not at our disposal now a picture of Europe which would permit us to make immediate decisions. New membership can take part only as part of an all-European effort including arrangements for and with Russia." Would Russia now have a direct ability to influence the process? Was NATO prepared to offer its security guarantee—that is, the stationing of its troops and the inclusion of these countries under the "last resort" nuclear umbrella? Would new members prove "politically reliable"? Would the U.S. Senate admit a country governed by ex-Communists? Who would fund a wider alliance? How would NATO enlargement, in contrast to EU membership, specifically promote internal stability in the new member (even though Greco-Turkish relations were commonly cited as an example of NATO's, particularly U.S., stabilizing influence on allies)? Even if the summit set out guidelines for eventual members, should they be published openly? Aside from recognition that the summit should send a "signal" on enlargement, there were "two views" or more within NATO on almost all of these questions.

Then came the PFP.

3

Partnership for Peace

The PFP did not spring from a single source. By late 1990 Polish academic and parliamentarian Wojciech Lamentowicz had already floated ideas for bilateral defense and security agreements between NATO and countries who would eventually leave the Warsaw Pact. These countries would enjoy "cross guarantees" while they remained in the Warsaw Pact political structure, with NATO emerging as the "core of the future all-European security arrangements."[1] On November 19 of that year the president of the Czech and Slovak Federal Republic Vaclav Havel called for NATO "association agreements" with European countries. Another root was the October 2, 1991, Baker-Genscher proposal for "joint action" with NATO liaison partners, whereas in early 1992 Romania began advancing ideas about "bilateral Partnership agreements" between NATO and its partner countries specifying duties, rights, forms and mechanisms for consultation and cooperation, as well as partners' opening individual missions to NATO. Recall also the prolonged U.S. effort to move the NACC into the operational peacekeeping domain, which it never did until PFP was agreed.

A Fresh Breeze from Evere

Nevertheless, the PFP as such originated in March–April 1993 as "Peace Partners" from then SACEUR John Shalikashvili's term "in lieu of new NATO membership" (which SACEUR did not favor for the near future). It was largely previewed in September 1992 by General James P. McCarthy, deputy commander-in-chief, U.S. European Command (CINCEUR):

> Beyond the need to stem the potential for crises, if NATO is to anchor the foundation of Europe's security for the future, the Alliance must re-focus its political-military capabilities on pan-European security concerns. Thus far, NATO has approached regional cooperation and the evolution of a new architecture with both lofty words and cautious actions. . . . If the NACC fulfills its potential, it will become the major forum of European security. However, without a willingness to engage significant security issues, *NACC risks becoming a facade.* . . . A fresh new breeze from Evere [NATO headquarters in Belgium] would bring an anticipatory approach to Europe's potential security concerns, especially those in the East. By stemming crises early and building confidence in a new security structure, Eastern Europe will more quickly develop democratic governments and open, prosperous markets."[2]

Specifically, General McCarthy proposed that NACC sponsor

- humanitarian relief planning and operations;
- NACC member-initiated security consultations with NATO "similar" to Article 4 of the North Atlantic Treaty (the right to consult should a threat arise);
- technical assistance for defense conversion, adequate militaries for national defense, and civil control of the military;
- military support for UN or CSCE political decisions;
- prepositioned peacekeeping equipment, mission planning and peacekeeping training seminars, the development of peacekeeping scenarios for NATO crisis management exercises; and
- a physical facility for cooperation "such as a NATO-NACC Annex near HQ NATO, or perhaps finding space for a 'cooperation wing' within NATO headquarters itself" (in December 1993 Belgium offered office space, which became the Manfred Wörner Cooperation Wing annexed to the NATO press building).[3]

Beginning in May 1993 the idea was then elaborated in the Department of Defense (DOD) by Assistant Secretary for Regional Security Affairs Charles Freeman, assisted by the principal draftsmen—Deputy Assistant Secretary for European and NATO Affairs Joseph Kruzel together with two other officials, Daryl Johnson and Clarence H. Juhl. Speaking at the opening of the Marshall Center in Garmisch, Germany, on June 5, 1993, Secretary of Defense Aspin called for consideration of "how our closer association with the democratic states of the East will evolve"; for the present, "innovative ways" should be found to draw them closer to the alliance, including "a permanent presence at NATO headquarters."

The first fully developed DOD paper was a draft of July 28, 1993, entitled "Concept Paper: Charter of Association with NATO." It was intended as an input to the interagency process being prepared for the forthcoming summit. The paper took final form on August 26, where the term "Partnership for Peace" first appeared following a conversation between Kruzel and Shalikashvili, and contained a draft "Agreement on a Euro-Atlantic Partnership for Peace"—the future PFP framework document, discussed below. DOD was particularly concerned to give NATO a new sense of purpose, to let the new European security order "define itself," while responding to Central European concerns without drawing "new dividing lines." Enlargement would have to be raised at the end, not the beginning, of a process of achieving interoperability with NATO and meeting alliance political standards, and new members would have to be "contributors" rather than merely "consumers" of allied security. In the DOD view, a PFP would also strike a better public chord than simply an upgraded NACC "work plan."

Other papers were being generated by the State Department and the National Security Council that were geared more toward "strengthening" the NACC, although Undersecretary of State Lynn Davis reportedly argued in an October 1993 memo to Secretary of State Warren Christopher that criteria for membership should be elaborated. Others placed Russia ahead of all else, or saw no reason for rush, the latter group said to be represented by Deputy Secretary of State Strobe Talbott and supported by DOD and the Joint Chiefs of Staff. Reportedly, Talbott argued in an October 19 memo to Christopher: "Laying down criteria could be quite provocative, and badly timed with what is going on in Russia."[4] Nevertheless, over the summer U.S. officials also said that "the initiative has to come from the Europeans." The views of Senator Richard Lugar (keen on new members) and Senator

Sam Nunn (concerned about whether NATO could make good on its promises) were also influential, whereas Secretary of State Warren Christopher and National Security Adviser Anthony Lake were believed to be moving in favor of NATO enlargement. As Senator Lugar revealed in December 1993 after the decision to adopt PFP:

> The proposal is somewhat of a disappointment to me in part because it was my understanding that thinking in the U.S. government had come a long way over the past three months and that expanded membership was coming to be seen as a function of the alliance's own strategic priorities. Various paths to expanded and full membership were being considered, whether those paths ran through an enhanced NACC or through intermediate stages such as associate membership. The State Department had developed a rather robust position on the question of expanded membership, and the PFP represented merely one alternative in the DOD bag of options. . . . In many respects, PFP epitomizes the Administration's *ad hoc* approach to European security problems. It is a band-aide offered in place of corrective surgery.[5]

In July 1994 this view was expressed somewhat more vividly by one U.S. official who was intimately involved in outreach activity from the very beginning (and who requested anonymity):

> No one thought it was really a good idea to go to the summit. A face-saving device was needed. The Alliance gave the impression that it was backing down after the [September 1993] Yeltsin letter. PFP was a non-policy to get everyone off the hook. It falls short of addressing a timetable and criteria. There is nothing behind it. It is pure [expletive deleted]. But there is an opening through which some states will force NATO to let them come in.

As far as actual enlargement went, however, the congressional testimony of August 12, 1993, of U.S. assistant secretary of state Stephen Oxman was not yet at the point of answering the administration's own questions:

> Are the Allies prepared to extend this security guarantee and commit their troops to defend the countries of Central and Eastern Europe? Whether the countries of Central and East-

ern Europe are or will be equipped to carry the burdens that membership in NATO entails? Will NATO work as effectively when it has several additional members whose consensus must be obtained before NATO can act? Would failure to admit Russia, while admitting the nations of Eastern Europe, isolate Russia from the new security system we are trying to build? . . . Would [Ukraine] want to be left as a buffer between Russia and a NATO extending to the Polish border? If satisfactory answers are found, then we can consider expansion of NATO.[6]

The Allies

As with most NATO initiatives, the United States was the prime mover behind the outreach and the attempts to inject NACC with substance. Again, during the summer of 1993 a U.S. NATO Mission diplomat stated that Washington wanted the Europeans to take the lead, but this did not happen—even though as early as May 21, 1993, at the NAA Spring Session in Berlin, German defense minister Volker Rühe squarely called for bringing the Visegrad into NATO:

> With their forthcoming association with the European Communities, the political foundations have been laid for the Visegrad states—Poland, Hungary, the Czech Republic, and the Slovakian Republic. I therefore see no reason in principle for denying future members of the European Union membership of NATO. To me, therefore, accession of new partners is not so much a question of 'if' as one of 'how' and 'when' [a phrase popularized the following January by President Clinton]. It is a question of timing and preconditions. This has to be determined.[7]

For example, as of October 1993 the UK view was "still being formulated." British defense secretary Malcolm Rifkind pursued the open-ended "no new lines" approach:

> While Poland, Hungary, the Czech Republic, and Slovakia will one day be able to enter the alliance, it is highly unlikely that the same thing could happen in the future for Belarus and Ukraine, and it is inconceivable that Russia could enter.

For the time being, our objective it to avoid creating new divisions in Europe which could encourage Russia to think that it has a free hand in the neighboring countries [excluded from NATO] that used to belong to the Soviet Union.[8]

But concern was also raised that if things went wrong in Russia it would be better for the West that Russia's neighbors not be part of NATO, as could be interpreted the remarks of David Gillmore, permanent undersecretary of state in the Foreign and Commonwealth Office: "If . . . Russia's neighbors become the targets of Moscow's renewed expansionary ambition, and if these neighbors had developed close relationships with the West and its organizations, then we could see a return to tension, even perhaps an element of confrontation in Europe."[9] The implication could have been that the West would be better off permitting a gray zone to its East than deterring adventurism by extending NATO security guarantees to Central Europe.

The signals from Germany were mixed. Foreign Minister Klaus Kinkel observed that "it must be possible to establish closer relations between NATO and these countries and to finally accept them as members to meet their security needs. Yet one must be careful and avoid new ruptures."[10] Defense Minister Rühe also pursued the parallel new NATO membership/Russia-NATO strategic partnership approach: "We cannot have stability in Europe without Russia, or even against Russia. That is why I have always said that an expansion of NATO, to include East European democracies, must be complemented by a strategic partnership between NATO and Russia."[11] But he also declared: "Preemptive crisis management for us Germans means that we move the Western stability zone as far as possible to the East."[12] Militarily, planners argued that the defense of Berlin requires Poland in NATO. However, although Bonn's reported original aim was to set "a date and conditions for Central European entry at the NATO Summit,"[13] by December 1993 Foreign Minister Kinkel conceded that for the time being NATO would offer agreements but no security guarantee.

The Russian understanding of the NATO's various national positions toward enlargement was assessed by the Russian Foreign Intelligence Service in *Prospects of NATO Expansion and Russia's Interests*, a November 1993 public document no doubt intended to discourage the Central Europeans:

- The United States had not yet formulated its position but was inclined toward the PFP to escape the "difficult situation" caused by enlargement appeals;
- The United Kingdom had not yet come to "a definitive appraisal" of the prospects for NATO enlargement, but was said to believe that a selective approach to enlargement was not feasible and to favor prior membership in the EU and WEU by way of a "disciplinary effect" on new NATO members;
- France believed any European state, not just the Visegrad countries, had the right to become a NATO member, such that no prospective candidates should be named and no specific dates for joining set, and NACC should not be given more powers to the detriment of the CSCE;
- Germany supported expansion as a matter of principle, giving preference to the Visegrad countries and extending the possibility of NATO membership prior to EU membership so as to make its own territory safe from instabilities in the region;
- The Netherlands was said to favor granting Article 4 rights to the Visegrad and offering a commitment to consider possible NATO action in the event of aggression, but putting EU and WEU membership first absent an increased threat from the East (Dutch officials corrected this analysis to note they favored parallel EU and NATO membership);
- Italy and Spain favored only preliminary consideration of expansion "because of Russia's painful reaction and the riskiness of granting NATO guarantees to countries whose foreign and domestic policy courses were not yet settled"; and
- Turkey, Greece, and Portugal were guarded about enlargement out of fear that aid would be diverted away from them.[14]

All in all, it seemed that the task of Russian propaganda, given the transparent indecision they perceived in the Western camp, would not prove inordinately difficult in blocking or slowing down enlargement.

Travemünde

On September 11, 1993, a conference was held at Truman Hall, the residence of the U.S. ambassador to NATO, among U.S. and alliance officials including Kruzel, Shalikashvili, Wörner, Freeman, Juhl, Commander-in-Chief Allied Forces Southern Europe Admiral Jeremy Boorda, and U.S. Mission NATO policy planning officer Captain Charles Dale. There the idea of Combined Joint Task Forces (CJTF), a rapid response and highly flexible structure that would allow for use by either NATO or WEU nations plus nonmember states of either organization, and the PFP were joined together. As a result, future PFP partners could participate not only in peacekeeping seminars but in actual military missions.

The "prevailing view" within the White House was settled in October 1993 by Strobe Talbott, supported by Aspin, following the Yeltsin letter. Speaking in Travemünde, Germany, where the Defense Planning Committee (DPC) met on October 20 and 21, Secretary Aspin revealed the concept open to "anybody in CSCE" as follows during the press conference:

> A country that wanted to be a member of NATO—or one that didn't—could join into a partnership agreement with NATO . . . each country that became a partner would prepare an implementation plan or a work program specifying the extent of the participation. And what those plans would include would be inclusion in military exercises and planning, and commitments to work towards the goals of transparency in defense budgeting, towards civilianization of its defense ministry and interoperability with NATO forces. . . . The notion here is that what we are trying to do would build an added capability to deal with things like search and rescue, disaster relief, peacekeeping, and crisis management. . . . All these people here would be under Article Four . . . We expect NATO membership to be increased, and I think most people thought that that meant sooner rather than later. But as for any specific date, I can't help you.[15]

Nearly a year later, Italian permanent representative to NATO Giovanni Jannuzzi recalled:

> The Allies first heard of the Partnership for Peace at the informal meeting of Defence Ministers in Travemünde in October 1993. . . . I remember the surprise, which lasted only briefly, and then the general feeling of support that emerged . . . everyone present understood that the problem was at the center of NATO's future and that the time had come to go beyond the NACC.[16]

The Clinton administration rationale for the PFP was to meet the long-declared security concerns of the Visegrad and other nations, to avoid destabilizing the situation in the former Soviet Union, and to preserve a functioning alliance. To satisfy all three requirements was admittedly difficult. Thus, in some ways the PFP was not altogether different from NACC: membership in NATO was held out, but involved no timetable, binding promises, or criteria; as an "NACC activity" and not a new institution, the PFP would be open to all on the same basis without preconditions (although, once a partner, the state would have to commit itself to the "goals" of transparency in defense budgeting, civilianization of the defense ministry, and interoperability).

At the same time, the PFP could be seen as *a path into NATO* for those who became "active" partners in choosing for themselves the extent of their engagement with NATO. Partners would "self-select" themselves, although the alliance would direct the process, as with NACC.

On December 3, 1993, Secretary Aspin described five "big advantages" for both allies and partners:

> First, it does not re-divide Europe. We spent two generations trying to lift the Iron Curtain. We don't want to replace it by drawing another line.
>
> Second, Partnership for Peace sets up the right incentives. In the old, Cold War world, NATO was an alliance created in response to an external threat. In the new, post-Cold War world, NATO can be an alliance based on shared values of democracy and the free market. Partnership for Peace rewards those who move in that direction.
>
> Third, Partnership for Peace requires that partners make a real contribution. It doesn't just ask what NATO can do for its new partners, it asks what the new partners can do for NATO. Security consultations, for instance, will be available to active partners [a new formulation, the prior being that NATO would consult with "states participating in the PFP"], those who make a contribution and involve themselves in the multinational activities that are the heart of NATO.
>
> Fourth, it keeps NATO at the center of European security concerns and thereby keeps American involvement at the center of Europe.
>
> Finally, it puts the question of NATO membership for the partners where it belongs, at the end of the process rather

than at the beginning. After we have some experience with the partnership process, it will be much clearer who among the eligible nations genuinely wants to buy into the NATO ideas of shared democratic values and cooperative security.[17]

In the run-up to the summit,"hundreds" of alliance man-hours would follow Travemünde in working out remaining details and sorting out an expanding list of around 50 questions. The questions ranged from first order issues, such as what precisely were the political and military objectives of the PFP and whether it was a substitute for or a pathway to membership, to legal contractual and financial issues that stemmed from a late-in-the-day initiative unveiled to allies only three months before the summit. That initiative was, moreover, astonishingly ahistorical in Aspin's contention that "NATO was an alliance created in response to an external threat" and that it was only for the future that NATO could be an alliance "based on shared values of democracy."[18] (The integrated military structure only came about after the launch in June 1950 of the Korean War, whereas the alliance was from the very beginning a community of shared values whose parties pledged, as stated in the preamble to the Washington Treaty, "to safeguard the freedom, common heritage and civilization of their peoples, founded on the principles of democracy, individual liberty and the rule of law.")

The reaction from the intended beneficiaries of PFP varied. Albania, Bulgaria, Romania—concerned about being consigned to some outer tier of cooperation—reacted favorably. Not surprisingly, President Yeltsin described it as "great" and pledged that Russia was "prepared to do its part in developing it."[19] The Visegrad nations were "realistic," and it was reported that "unlike Budapest and Warsaw, Prague is not pressing for explicit security guarantees from NATO in the interim" but rather envisaged a "sort of *political association* . . . which could begin after the summit" followed after "several years" by membership, with the Czech Chief of General Staff having stated that full integration militarily would not be possible before 2000 [emphasis added].[20] Nevertheless, according to President Havel in a statement before the Chamber of Deputies on October 12, 1993, the very fact that many wars had either begun or ended in his country provided a historical lesson, as did the capitulation of Munich, that the Czech lands could not look after themselves. They would have to engage in all-European activity and be firmly incorporated into a functioning collective defense system that embodied Euro-American civilization.

The Poles were the most adamant about keeping NATO's feet to the fire. During a December 3, 1993, NACC meeting, Polish foreign minister Olechowski stated: "We expect, particularly from the Allies, a greater understanding for our views on security" and "a more positive response to our needs, aspirations and contribution that we want and can bring to NATO." However, no consensus was found at the ministerial meeting on a Polish proposal to include language in the communiqué declaring the "open nature" of the alliance—reflected also in the statement by Bulgarian foreign minister Stanislav Daskalov that "Bulgaria shares the justified expectation that the Summit will send a clear political signal of NATO's readiness to open to new members." Instead, the statement merely noted "the need for opening new perspectives for the consolidation of stability and security in our region, keeping in mind the aspirations of all NACC member countries." A Bulgarian proposal was not accepted, however, that would have specifically included *equipment interoperability* and *standardization* in the new NACC Work Plan and would have been a move toward additional practical military steps.

The Polish delegation had considered placing a footnote in the text to register dismay about the NACC's failing to issue a sufficiently strong signal to the summit. In the end, however, Foreign Minister Olechowski stated that the PFP would meet Poland's expectations if, *inter alia*, its consultative provisions could lead to "the definition of joint actions aimed at restoring and maintaining peace and security" (suggesting an interim security guarantee?), if bilateral arrangements would allow for diversification according to the partner's interest and take into account progress in the development of democratic institutions, and if all partners would be presented with identical preconditions, so that one country's progress would not be blocked by other countries.

On December 9, 1993, Olechowski and his Czech counterpart Jozef Zieleniec appealed for the summit to send a clear signal that the alliance would enlarge to Central and Eastern Europe. During the NACC meeting, a similar message was heard from Estonia, Lithuania, Bulgaria, Romania, Slovakia, Hungary, Albania, and Latvia—10 aspirants in all.

4

The Brussels Summit

The Partnership for Peace as adopted at the January 10–11, 1994, NATO summit in Brussels more or less resembled the Aspin plan. The PFP was described as a "new security relationship" intended, much like NATO membership, to increase stability, diminish threats to peace, and build strengthened relationships by promoting practical cooperation and commitment to democratic principles. Partners would be invited to send permanent liaison officers to NATO and to a separate Partnership Coordination Cell (PCC) adjacent to SHAPE. They would also commit themselves to five objectives:

- transparent national defense planning and budgeting;
- democratic control of the armed forces;
- contribution to UN or OSCE operations;
- development of cooperative military relations with NATO for undertaking peacekeeping, search and rescue, humanitarian operations, "and others as may be subsequently agreed," that would include access to relevant NATO technical data; and
- development over the longer term of interoperability with NATO nation forces.

In addition, NATO would consult with any active that partner perceived a direct threat to its terri political independence, or security. The PFP di Aspin plan, however, in at least six ways, the sa only time will measure.

First, no reference was made to "crisis management." An explicit mention presumably would have suggested partner involvement in the NATO Precautionary System for political or military responses to the likely challenges to security as opposed to deterrence of and defense against the Warsaw Pact threat as in the past—that is, access to the NATO information and intelligence Situation Center and communications center and participation in exercises of a degree beyond peacekeeping. France was said to have argued that crisis management was within the competence of the (then) CSCE (although the CSCE has no "enforcement" mandate, and crisis management explicitly featured in the 1991 NATO Strategic Concept, which France participated in elaborating). Instead, the invitation to join speaks of "creating an ability to operate with NATO forces in such fields as peacekeeping, search and rescue and humanitarian operations, and others as may be agreed," with peacekeeping field exercises—which proved so elusive over the years within the NACC framework—proposed to begin in 1994. It is, thus, arguably more open-ended than the original formula, and, in fact, crisis management did feature in the first PFP work program (briefings and an exercise). Moreover, the summit endorsed partner participation in CJTFs, which could cover both collective defense and out-of-area missions as in Bosnia-Herzegovina.

Second, the language on enlargement could be interpreted as either a clear declaration that the alliance would take on new members or as too helpful to those who sought delay: "We expect and would welcome NATO expansion that would reach out to democratic states to our East, as part of an evolutionary process, taking into account political and security developments in the whole of Europe." On January 8, 1994, President Clinton conceded that there was no consensus to expand NATO, and that the alliance did not wish to appear as drawing "another dividing line in Europe"—whatever that was supposed to mean; at the same time, he declared on January 12 in Prague that "the question is no longer whether NATO will take on new members, but when and how." It was conceded by a high-ranking SHAPE officer at an NAA meeting in February 1994 that there were still "different interpretations" of the PFP among the NATO countries: "those who believe PFP is a gateway to full membership in the near future, and those who do not believe in the wisdom and perspective of full membership shortly." It would thus be "premature" to make predictions.

Put another way, as a Canadian diplomat recalled, the "PFP helped us get out of the hole we were digging on expanding NATO."

At least some partners were relaxed about the way ahead, predicting that the *gemeinschaft* forged from the intermingling of diplomats and officials at NATO would eventually obscure barriers between members and partners. Nevertheless, if Russia were always NATO's major preoccupation, would there ever be a good time to enlarge? On June 18, 1993, Admiral William Smith, U.S. military representative to NATO, was candid enough on this score: "We must be careful to prevent a new division from taking hold . . . that separates Western and Central Europe from the former Soviet Union. On this argument, what happens in Poland and Hungary, for example, is important to us; but what happens in Russia is crucial."[1]

The Russia factor also confused the issue as to *why* NATO would expand—again, in response to a threat as during the Cold War, or in pursuit of democratic integration as the founders envisaged—and this difficulty would stretch out for many months. For example, on February 1, 1994, Assistant Secretary Oxman presented senators with this formula: NATO should not "foreclose the best possible future for Europe: a democratic Russia committed to and working with and for the security of all its European neighbors. . . . But, at the same time, [PFP] preserves the means to deal with a darker future, should it occur."[2]

Why, however, was there an apparent presumption that NATO enlargement and cooperation with Russia were incompatible or, at least, had to be treated together? What was the definition of the mantric "best possible future?" And was not the idea that NATO would expand rapidly if and when Russia became a new threat incredible? Moreover, although Ambassador Robert Hunter, the U.S. permanent representative, was quick to remind that "none of the aspirants for joining NATO are really threatened now," neither were the NATO allies. So why was the threat environment considered a factor in NATO membership?

Of course, even if NATO did expand there would be no certainty that North America and Western Europe would defend new members, just as the German-Soviet partition of Poland in

1939 hardly sparked a robust Anglo-French response despite treaty commitments. But this was true for present members as well, because the Washington Treaty in fact contains no ironclad security guarantee: Article 5 simply commits each member to take "such action as it deems necessary, including the use of armed force, to restore and maintain the security of the North Atlantic area." (Compare the 1954 WEU Treaty of Brussels, which provides that members "will . . . afford the Party so attacked all the military and other aid and assistance in their power"). NATO membership, however, could at least offer the same "guarantee" other allies enjoyed and thus transform a vacuum into a reasonable expectation. It would offer political reassurance in security terms and, sometimes overlooked, would provide confidence to foreign investors—themes that General McCarthy evoked more than a year before the summit.

Nevertheless, the PFP invitation also spoke of promoting "closer military cooperation and interoperability"—a key requirement for membership in NATO. This began with the release of an unclassified manual on peacekeeping "exercises and tactics" and two NATO Standardization Agreements (STANAGs) on quality and quality assurance in defense procurement contracting and technical documents relating to acquisition practice, quality assurance, and matériel in April 1994. The NATO Conference of National Armaments Directors (CNAD) began early that year to sponsor multinational expert teams from NATO nations related to defense procurement management in response to requests for expert advice from Hungary, Bulgaria, and Poland.

Third, although the PFP invitation was supposed to be a communication from the NATO secretary general to other nations, it was published in a generic, public form because the NAC might not agree on which countries to invite (for example, Cyprus and the "former Yugoslav Republic of Macedonia"). The phrase "other CSCE countries able and willing to contribute to this program" was intended to put "certain markers" around states likely to cause controversy.

Fourth, Article 4 of the North Atlantic Treaty governing consulations did not explicitly feature in the summit PFP documents because a number of allies were concerned that this would imply political membership of NATO. But even here some debate took place. During the December 8, 1993, meeting of the NATO DPC,

it was reported that "some ministers . . . suggested the consultation rights should be phased in rather than granted at the start,"
with a German diplomat quoted as suggesting that consultations
should be accorded only after a partner program was well under
way and "has proved its worth" to NATO. France reportedly
opposed consultations of any kind.[3] In addition, a partner consultation request would not necessarily be granted, unlike General
McCarthy's reference to their being "[NACC] member initiated";
a partner consultation request, according to U.S. secretary of defense William Perry on May 24, 1994, "could" only involve a
convening of the North Atlantic Council to "consider" the request
by the partner. These consultations, morever, would be conducted only with "active" partners. Thus, a partner would have
to jump two hurdles—NATO's consent and NATO's appraisal of
the country as an "active" partner—to obtain consultations from
NATO, even though the invitation clearly stated that "NATO
will consult with any active participant in the partnership if that
partner perceives a direct threat to its territorial integrity, political
independence, or security."

Fifth, the U.S. approach of offering PFP without preconditions to any CSCE state could be said to have been modified,
at least *pro forma*, in favor of paragraph two of the framework
document, which states that "protection and promotion of fundamental freedoms and human rights, and safeguarding of freedom, justice, and peace through democracy are shared values
fundamental to the Partnership." It also calls for adherence to
arms control commitments and other international obligations
and highlights three generic provisions of the UN Charter: to
refrain from the threat or use of force against the territorial integrity or independence of any state, to respect existing borders, and
to settle disputes by peaceful means.

So general was this that Poland thought a more detailed enu-
meration should have been offered. Yet, in Naples on July 8,
1994, President Clinton claimed, astoundingly, that the *mere
pledge* to refrain from the threat or use of force meant that Russia
had "recognized the integrity of the borders of its neighbors."
On May 31, Secretary Christopher even stated that this pledge
amounted to "certain security guarantees."[4]

Nevertheless, Secretary Christopher stated on May 31, 1994,
that there could be no specific and measurable criteria, and that

the primary consideration was political: the question of whether benefits would flow in both directions. Partners sought specific criteria, but too detailed an approach could have limited alliance flexibility: a state could at some point come before the NAC to say it had complied with detailed criteria, yet there might be reasons for the alliance to defer its admission (for example, the aspirant's version of how it had settled minority questions could be disputed by its neighbors).

Sixth, although the issue was still being debated in NATO's Special Political Committee the week before, the partners would be expected to foot their own bill. At the December 1993 DPC meeting, ministers reportedly were split on how to finance the NATO end of the partnership activities, with Belgium insisting that funding come from existing budgets. U.S. principal deputy under secretary of defense Walter Slocombe on February 2, 1994, clarified that the PFP was not an assistance program and that most allied and partner costs would be paid directly from national budgets. However, some shared costs from the NATO common budgets—civil, military, and infrastructure—would require start-up funding.

The NATO common budget would pay an estimated $4 million to $8 million (from a NATO total, according to 1992 figures, of $2.06 billion including the pipeline and administrative budgets). Initial estimates indicated, however, that start-up costs for 1994 would amount to $10–14 million, covering establishing partnership offices at NATO headquarters and SHAPE (Building 104, the prior "Live Oak" facility dedicated to planning the defense of Berlin, as temporary headquarters prior to the move to Building 901), field training and exercises (most of these are national costs), and training courses. The U.S. government would contribute about 25 percent of these costs.

On May 24, 1994, Secretary Perry responded to partner concerns that, at least for those embarked on their way into NATO, the scale and costs of integration would require more than the limited resources partners could contribute on their own. He informed the DPC that the issues of adequate PFP funding and the drawing from NATO common budgets needed to be examined, as did the need for better coordination among bilateral security assistance programs. Perry also urged that at least a basic defense planning review begin before the end of the year.

Presentation

On January 11, 1994, the NACC representatives as well as the "neutral" NACC observers and ad hoc group participants were briefed on the PFP. It was noted that the alliance had gone beyond a simple reaffirmation of Article 10—suggesting that this may have been an option—but had fully committed itself to enlargement. As German defense minister Rühe revealed two days after the summit: "Partnership for Peace is not a substitute for opening up the alliance, and what was said in Brussels was only achieved after months of effort. It was not planned that way at the outset; in fact, to begin with it was seen as being more of a substitute for membership," as had been the SHAPE intention.[5]

Procedurally, partners were requested to organize their "presentation documents" into two sections. The first would identify resources and set forth cooperative activities of interest, assets that might be made available for PFP activities (infrastructure, logistics, and medical support), and long-range plans (including force development goals and research and development). The second would cover the steps to be undertaken to meet PFP political objectives (adherence to democratic principles, good neighbor relations). NATO and the partner would then discuss the contents and agree upon an Individual Partnership Program (IPP), which would be made transparent to other partners (but not made public unless the partner chose to do so, as Hungary did).

On the political level, an NACC/PFP forum would address general policy matters, and the NAC would also meet in an "NAC + 1" format to discuss issues with an individual partner or in an "NAC + n" format with partners engaging in a specific activity (for example, the optional defense planning review). At the working level, a "Politico-Military Steering Committee" (PMSC) would meet individually with partners, in NACC/PFP format to consider the more specific Partnership Work Program and merging with the prior NACC Ad Hoc Group on Cooperation in Peacekeeping, and at "16 + active partners" for defense planning. A "Military Committee in Cooperation Session" would meet with all or some partners, and partners would be invited to send permanent liaison officers to NATO headquarters and to the PCC where military planning would be carried out.

Poland presented the first document on April 25, 1994.

Twelve joint activities were suggested, with air defense and command, control, communications, countermeasures, intelligence and information systems assigned as "priority areas." Brigadier General Stanislaw Wozniak did not appear uncomfortable with the initial focus on peacekeeping, as he informed an NAA-NATO-Polish Sejm seminar on April 18, 1994, in Warsaw:

> The subject of joint training and exercise planning proposed by NATO does not seem very appropriate at first sight. However, in order to carry out any joint action, either rescue missions or peace operations, cooperation between military units and headquarters is necessary. The units must speak the same language, use the same codes, compatible telecommunications systems, compatible bases and, if possible, similar equipment. Moreover, in the case of 'peace enforcement' operations [which could read "defense of Poland"] the preparation of units for joint combat is necessary; practically speaking, this entails the need for standardization of combat equipment and munition [and] appropriate logistic support. . . . Thus, these slogans that do not appear to mean very much represent a huge potential and their utilization depends to a large extent on the both sides.[6]

Over the longer term, the Polish military was interested in establishing liaison missions not just at SHAPE and NATO but at Major Subordinate and Principal Commands and key units, including the ACE Rapid Reaction Corps (ARRC). Another interest was participation in the "combined joint task forces" (CJTFs) decided in principle at the January 1994 summit to allow for NATO or WEU "coalitions of the willing" to perform collective defense or crisis management and peacekeeping and peace implementation missions. The idea of establishing a NATO mission in Poland was also raised. For two years Poland had proposed NATO peacekeeping facilities on its soil (the "too close" cooperation that some NATO nations sought to avoid in the belief that such "differentiation" would provoke Russia).

The idea of Polish missions at NATO Commands or NATO missions in Poland were far ahead of the PFP menu. So too was the proposal during the aforementioned Warsaw seminar of Andrzej Towpik, director of the Department of European Institutions in the Ministry of Foreign Affairs, for *associate membership* and *de facto* security guarantees. Nevertheless, at least U.S. officials urged partners to be ambitious, noting that NATO was in "uncharted waters."

The Romanian presentation document was submitted three days later. At an NAA seminar in Sinaia on July 14, 1994, it was described by Defense Minister Gheorghe Tinca as proceeding from certain advantages his nation had accumulated as a result of its partial break from the Warsaw Pact in 1967: regulations and instructions that were not under Soviet influence and compatibility with Western systems in some areas (for example, the civilian side of Romanian air defense had long employed Western technology over Russian technology).

Although the Polish presentation document was a nononsense, largely military text that took NATO at its word, Hungary stressed consultations and discussion of regional security issues—no doubt because of anxieties regarding NATO operations in neighboring former Yugoslavia that took advantage of Hungarian airspace. (Since the conflict erupted, Hungary had been seeking at least provisional security guarantees; it was concerned, among other things, about an attack against its nuclear power stations. Some Hungarian diplomats, however, felt compelled to point to this risk because, they said, NATO officials had told them that there was no need to expedite enlargement in the absence of a threat.) Hungary was the first country to apply for full NAA membership and to provide contributions to the NAA-wide budget.

Bulgaria's ex-communist government was at odds with reformist and pro-NATO president Zhelyu Zhelev and did not apply for NATO membership until 1997. Bulgaria was also interested in mutual consultations and expressed particular attention to reciprocity and dialogue before decisions were taken. Because it had little interest in peacekeeping and in any case could not, as a neighboring state, participate in former Yugoslavia, its priority was getting on with nuts-and-bolts issues such as access to interoperability information. Both the Czech and Bulgarian approaches included interest in reorienting their arms industry, including coproduction of weapons with alliance nations configured to NATO standards.

Although not seeking membership in NATO, Finland, on the other hand, submitted its presentation document on May 9, 1994, out of concern about its own neutrality. It specifically noted that the PFP initiative "is aimed at the countries of Central and Eastern Europe." Finland would not be seeking, however, through PFP change its policy of "military non-alliance and independent defense." Sweden also emphasized nonparticipation in military alliances albeit recognizing the value of NATO in peacekeeping missions. Here the PFP could, nevertheless, positively influence the attitudes of these and other neutral and nonaligned nations toward joining the EU and eventually framing a common defense policy, which would have to exclude neutrality.

Aftermath

Although the PFP framework documents were rapidly signed, the post-Travemünde reactions—relief, complacency, frustration—from prospective partners paralleled post-summit reaction. Polish foreign minister Olechowski considered that because no timetable was defined, the PFP amounted to a Polish diplomatic "failure."[7] A Polish diplomat argued that "there is no balance, the concern is always with Russia." In Prague on January 12, 1994, President Clinton, and by implication the alliance as a whole, was more or less scolded by President Walesa:

> It is difficult for me to hide my doubts and reservations. . . . The idea of a divided, confrontational Europe has revived. There is Russia, who, perceiving NATO as an enemy, fears an increase in its membership, and is resorting to blackmail. There is the West, which is vacillating. We are reproached with distrust, distrust in the face of agreements and treaties. Poland's historical experience, and not only historical, in this respect has not been good. We quite recently signed a declaration concerning the question of Poland's entry into NATO [the 25 August 1993 Polish-Russian Joint declaration]. And what of it? What kind of force did it have? Scarely a couple of weeks or so have passed. Is it surprising that we are distrustful? We have learned to believe only facts, for it is facts which create guarantees.[8]

Deputy defense minister Jerzy Milewski claimed that not all NATO members commanded "a full understanding as to why NATO should be expanded at all." He believed that the main opposition came from Spain, Portugal, and Greece to prevent cuts in NATO assistance, whereas the "most understanding was exhibited by Germany, Denmark, Norway, and Turkey."[9] It should be recalled, of course, that NATO nation attitudes toward enlargement were also conditioned on geopolitics: integrating Bulgaria, Romania, and Albania was important for Istanbul; for Bonn and the Hague the Visegrad nations had priority.

In contrast to President Walesa, Russian foreign minister Kozyrev called the PFP a step "in the right direction . . . something which essentially is based also on our proposals to the effect that now, instead of a quantitative membership expansion eastwards of NATO, one should be occupied with the practical aspects of interaction," mentioning joint maneuvers and peacekeeping operations in conflict areas. He did not rule out that Russia "will decide, in the end, to join NATO." It was also conceivable, he said, that NATO would merge "into some broader European structure."[10] *Izvestiya*, however, crowed: "Russia's vigorous diplomacy was crowned with the utter humiliation of East Europe and the Baltic states. . . . mindful of Russia's interest, the Americans devised Partnership for Peace, a vacuous program which binds no one to anything."[11]

Despite these articulated Polish doubts about the PFP's being a pathway into NATO and Russia's satisfaction that it was not, President Clinton indirectly suggested in Prague on January 12 that even before NATO expanded, neither the United States nor NATO would stand by as it had in 1956 or 1968: "Let me be absolutely clear: the security of your states is important to the security of the United States."

The Brussels summit decisions were hardly received with uniform equanimity by the political elite in the United States. As former White House deputy chief of staff in the Bush administration Robert Zoellick stated, "The problem is that there has been no strong counterbalancing force making the case for a European policy separate from Washington's Russian calculations." The PFP, he continued, "does not appear to offer anything beyond NATO's 1991 decision" to create the NACC (for example, it could be viewed as simply the next stage of peacekeeping cooperation). Instead, the United States should propose criteria that, if met, would enable "at least the Poles, Czechs and Hungarians to qualify for NATO membership over the course of about three to six years."[12]

Henry Kissinger identified what he saw as a fundamental problem with the Clinton administration: "This tendency toward abstraction is compounded by an extraordinary obsession with public relations . . . treating foreign policy as if it were a domestic issue susceptible to consensus through trade-offs."[13] He urged that NATO face the fact that "some form of Visegrad membership" is inevitable and that Russia could be given assurances that no foreign troops would be stationed on the soil of new NATO members (as with the former GDR under the 1990 Treaty on the Final Settlement of Germany). But he considered the PFP a "vague, multilateral entity specializing in missions having next to nothing to do with realistic military tasks" and equating "the victims of Soviet and Russian imperialism with its perpetrators." Kissinger concluded: "If the Partnership for Peace is designed to propitiate Russia, it cannot also serve as a way station into NATO, especially as the administration has embraced the proposition rejected by all its predecessors over the last 40 years—that NATO is a potential threat to Russia."[14]

Zbigniew Brzezinski argued that the PFP was designed to avoid jeopardizing U.S.-Russian relations. In exchange for *a formal treaty of NATO-Russian alliance*, however, a simultaneous initiative could be made to establish a "NATO-linked coalition for regional security with the three or four Central European states qualifying for eventual NATO membership." But he warned against accepting any Russian designs whereby Central Europe would be kept out of NATO and be recognized as a zone of particular Russian interest.[15]

Professors Kissinger and Brzezinski both lent their support to the "NATO Expansion Act of 1994" sponsored on April 14, 1994, by Congressman Benjamin Gilman, then ranking Republican and future chairman of the Foreign Affairs (later International Relations) Committee and a member of the House delegation to the NAA. The act declared the sense of Congress that Poland, Hungary, the Czech Republic, and Slovakia should be the first, but not necessarily the last, to be made full NATO members by 1999. In addition, the bill authorized the president to provide transition assistance to these and possibly other countries to facilitate their integration into NATO. The legislation was scuttled by the Rules Committee presumably on grounds that the bill, amending the Defense Authorization Act, had bearing on foreign assistance administered by the Department of State; most likely, according to those directly involved, the true reason was that the Democrats sought to avoid embarrassing the president.

Likewise, another Republican, Congressman Henry Hyde, on May 5, 1994, introduced the "NATO Revitalization Act." It urged NATO "to establish benchmarks and a timetable for eventual membership for selected countries in transition," which the bill cited as not only the Visegrad but the Baltic states as well. In his floor remarks, Congressman Hyde explained that the legislation "extends beyond the limited scope [of the PFP] without prejudicing its merits." He argued that "a new NATO that includes former Warsaw Pact members would be no more a threat to Russia than the old NATO—which was, and will remain, a defensive alliance."

On January 27, 1994, the Senate passed, by 94 to 3, a "sense of the Senate" resolution calling upon the U.S. government to "urge prompt admission to NATO for those nations after they have demonstrated such capability and willingness . . . to support collective defense requirements and established democratic practices including free, fair elections, civilian control of military institutions, respect for territorial integrity and the individual liberties of its citizens," as well as sharing the goals of NATO. This was a compelling argument against the case increasingly being made by U.S. officials after the summit that one of the reasons for offering the PFP and not new membership was that the support of the 16 parliaments of the NATO nations would not be obtained for enlarging NATO. Others would argue, however, that these political expressions fell short of fully considering whether the United States was prepared to extend the security guarantee, including its nuclear dimension, further East in Europe.

Senator Richard Lugar slung another arrow at the PFP on June 28, 1994, noting that the administration had rejected his approach of establishing a "clear timetable for the expansion of NATO to include the Visegrad countries." He termed the PFP a "policy for postponement" that might allow the EU or even Russia to dictate NATO enlargement. Instead, the PFP should begin with the premise of strategic differentiation, and the alliance should commit itself to admitting the Visegrad countries by 1998. He also urged that $40 million or so be allocated to finance U.S.-Polish joint staff talks, technical exchanges, officer training, and bilateral training and exercises. "The West must make plain to Russia that we are now discussing *how* to expand NATO, not *whether* to expand it. The best way to communicate that message is by agreeing on *a schedule for associate and then full membership* for the Visegrad states."[16]

On July 14, 1994, the Senate adopted an amendment offered by Republican senator John McCain urging NATO to invite the Visegrad countries to accede "at such time as each is in a position to further the principles of the Treaty and contribute to the security of the North Atlantic area" and called upon the president to urge NATO to declare criteria and timetables for new members and report on that progress 90 days after the passage of the legislations. The next day the Senate adopted, 74 to 22, a bill entitled the "NATO Participation Act" moved by Republican senator Hank Brown and Democratic senator Paul Simon. It authorized the president to transfer excess defense articles to "Poland, Hungary, and the Czech Republic, and declared that it is the sense of the Congress that these three countries should "be included in all activities . . . related to the increased standardization and enhanced interoperability of equipment and weapons systems."

NAA secretary general Peter Corterier, former German minister of state for foreign affairs under Helmut Schmidt, and many NAA members also contributed to the public debate on PFP. In a graduation speech at the NATO Defense College in Rome on February 11, 1994, Corterier declared:

> I am deeply disappointed that our present political leaders have made a strategic, but not irreversible, error. . . . Reformers in Central and Eastern Europe have staked their political careers on quick integration into Western institutions. If they are constantly rebuffed, including by the EU and its glacial process of offering even modest trade concessions, the reformers may eventually lose their political mandate. Progressive forces will be swept away. . . . As a German, I must recall that when President Gorbachev demanded that a united Germany not be part of NATO, even though that was in the Soviet Union's own direct interest, President Bush stood firm and we succeeded.[17]

WEU then "upped the ante" when, on May 9, 1994, the WEU Council of Ministers issued the "Kirchberg Declaration" creating a "Status of Association" for Bulgaria, the Czech Republic, Estonia, Hungary, Latvia, Lithuania, Poland, Romania, and Slovakia. Akin to prior proposals for NATO associate membership, these nations could henceforth directly participate in council discussions and by invitation in WEU working groups, carry on a liaison arrangement with the WEU Planning Cell, and engage not only in exercises but operations.

Even granted the different nature of the WEU, or the cynical view that participation in WEU operations was simply making use of "cheap labor without a vote," the difference with PFP was immediately clear: differentiation, association rather than partnership, the ability to take part in decisions, consultations on any subject, preparing not just for exercises but operations, and clearing a path toward full membership based on the logic of their entry into the EU with the WEU as its defense arm (the status only applied to those who had signed a "Europe Agreement" with the EU, an accord that provided political and technical criteria and a rough deadline for full membership—exactly what NATO was unwilling to do). Moreover, the decision was taken despite Russian discomfort, and the NACC-like Forum for Consultation was suspended in favor of associate partners directly participating in every other council meeting (every second Tuesday).

Subsequently, WEU and its partners participated in a unique collective drafting exercise much appreciated by partners, who felt that way and not as guests, published in November 1995 under the title *European Security: A Common Concept of the 27 WEU Countries*. This gave expression to what van Eekelen termed "security through integration" rather than "security guarantees" as the answer to the problems of the region. The first meeting "at 24" took place on May 27, 1994.

Although Russia was not part of this integration, WEU agreed to a trilateral arrangement whereby the WEU council chairman, secretary general, and representatives of the non-partner country could meet albeit not in any institutionalized form. This reportedly was the model which Russia sought to conclude with NATO.

Although the U.S. administration was said to have welcomed the WEU decision, in June 1994 Deputy Assistant Secretary of Defense Kruzel, during a meeting with NAA Political Committee members at the Pentagon, urged that WEU and NATO needed to coordinate their outreach programs. He recalled that when France and Germany had proposed WEU associate

status for EU associates this had confused the intended beneficiaries as to whether a "choice" was being put to them between
WEU or NATO. He welcomed the WEU decision as "another way
of showing them that they are coming into our institutions," but
cautioned that "it would be a strategic mistake to encourage the
impression that WEU is a new anti-Russian alliance."

5

A Special Partnership

Russian presidential adviser Andranik Migranyan questioned whether the PFP was actually aimed at "restraining and disciplining Russia itself" without Russia's "having a full voice in NATO."[1] Presidential adviser Sergei Karaganov, however, portrayed the PFP as a success for Russian diplomacy: "In the end, Russia's cooperation with forces in NATO countries which do not want the Alliance to be expanded seems to have averted the inevitable."[2] On the day after the summit, the Russian Foreign Ministry likewise described the PFP as "an alternative to a mechanical NATO enlargement and to an accelerated admission of the new members." The ministry would now hold consultations on the results of the Brussels NATO meeting with a large number of European countries to identify optimal ways of establishing "a genuine pan-European security system, common and collective response to any new stability risks on the continent, overall consolidation of the capacity of the CSCE [OSCE] under whose aegis other European structures—the EC, NATO, NACC, WEU, and CIS—might operate."[3] In 1992 NATO had declared it would consider supporting implementation of UNSC resolutions and the narrower option of CSCE peacekeeping, but the alliance remained wary of the apparent Russian effort to have CSCE control NATO, which would give non-NATO nations a veto on alliance freedom of action.

A NATO PFP "briefing mission" visited Moscow from February 28 through March 1, 1994. "We have no particular problems with this outline document," deputy Russian foreign minister

Vitali Churkin stated. In contrast to the NATO view that the PFP offered identical conditions to all prospective partners, however, he pointed out that Russia considered it a draft open to amendment: "It is not just a matter of signing the document but of inserting into it some specific thoughts on how we, Russia, visualize political and purely military partnership, specific applied partnership, with the North Atlantic alliance," codified in something broader than the PFP offer (recall that President Yeltsin referred to "developing" the PFP).[4]

This readiness to negotiate was, of course, a far cry from pre-perestroika propaganda that had described NATO as "designed to serve as an instrument of aggression and a means of pressure against the Soviet Union and the other socialist states."[5] Yet Russia clearly was not comfortable with simply accepting NATO's offer unconditionally. Russian diplomats actually viewed the PFP as not offering Moscow very much. For example, Russia could address its own security concerns, judging the PFP threat-based consultation mechanism as either irrelevant or too narrow. (Other partners such as Hungary and Bulgaria also sought a broader and bilateral consultative relationship with NATO.) Russia instead sought consultations on the broader agenda of European and global security issues. It also no doubt sought to enhance the international recognition of Russia, despite its massive transitional difficulties, as a Great Power, or *Derzhava*; cooperation with NATO should be "virtually as between Allies," and Russia should be viewed as an "equal partner," the NATO PFP briefing mission to Moscow was informed. For NATO, however, the whole idea of "strategic partnership" was not "very well developed at this point," although a "special PFP program for Russia" was reportedly not ruled out altogether at that stage.[6]

The Russian position remained ambiguous over the next several months. Defense Minister Grachev stated on March 18 that Russia would be ready to join by the end of the month. On March 31, presidential spokesman Vyacheslav Kostikov predicted a final decision in "six or seven months." The reasons he gave in favor of joining the PFP were to spur modernization and military reform, obtain a greater opportunity to influence NATO's political decision making (such as NATO's air operations over Bosnia), and secure a partnership between NATO and the CIS. On April 5, Churkin stated that Russia would sign the framework document on April 21, but this date was never officially suggested to NATO.

Then, on April 10 and 11, 1994, NATO conducted limited air strikes against Serbian military targets. Russia protested over what it argued was a lack of alliance consultation with Moscow that went against the spirit of the PFP, with NATO arguing that no further consultation had been formally required as the strikes had been undertaken under UNSC authority. On April 14 the Duma condemned "unilateral NATO acts of force," and on the same day Kozyrev declared:

> We are getting the impression that our Partners in the West, especially within NATO, have not found the correct solutions yet. We are interested in a much more serious approach, not just discussing some framework document. We also have proposals being worked out by the Russian Foreign and Defense Ministries, but we are interested in much more serious relations, which would rule out surprises and unilateral measures, especially military ones, in those areas where we should be cooperating more closely.[7]

The PFP was not a legally binding agreement, but the Duma demanded that it have a say. Different perspectives had been raised. On March 17, the chairman of the Committee for CIS Affairs and Links with Compatriots, Konstantin Zatulin, warned that the PFP would end Russia's "special role" as a guarantor of peace and stability in the CIS, and that it would compel deference to NATO. International Affairs Committee chairman Vladimir Lukin argued that PFP should have been jointly proposed by the United States and Russia and suggested, bizarrely, that "Russia joining this program is like when a rapist, having cornered a girl, gives her a choice: Either she can just give in or he will have her anyway. It turns out that on top of that the girl is supposed to pay."[8] On April 18, however, Defense Committee Chairman Sergei Yushenkov backed PFP as an alternative to a return to the Cold War, arguing that nothing in the PFP had not already been proposed in some form by the USSR or Russia.[9]

During an April 25, 1994, meeting in Moscow among officials of NATO, NAA, the Russian Foreign and Defense Ministry, and the Federal Assembly, the head of the Foreign Ministry NATO Relations section Andrei Androsov declared that "our attitude is positive" but that the "PFP should take account of our nuclear status and provide a well-defined mechanism of cooperation.

One framework document will scarely satisfy the level of cooperation we require." Sergei Rogov, deputy director of the Institute
of the USA and Canada, asked: "What was the mechanism for
decisions and how did Russia fit in," could PFP destroy the CIS
collective security treaty, and what would be the impact of NATO
standardization on the former Soviet armed forces and defense
industry? The latter issue was also a favorite theme of the upper
chamber, the Federation Council, considering its links to the
"military-industrial complex." The vice chairman of the CIS Committee, Yuri Voevoda, asked whether "Russia [might] become a
satellite of NATO" and whether China might perceive Russia as
the "front line" of NATO.

NATO assistant secretary von Moltke responded that NATO
and its PFP partners would not themselves make the fundamental
decisions about peacekeeping, which would come only from the
UNSC or the OSCE. Decisions about implementation would be
conducted in a "16 + 1 or 16 + n format, and in the Coordination
Cell we would decide what standards we could use." What Russia may have had in mind, however, was a *veto* right on NATO
operations outside the treaty area, whether or not Russia decided
to participate in an operation—a kind of UNSC approach to Europe.

On May 19, 1994, the deputy secretary of the Russian Security Council, Valeri Manilov, resurrected the old rumor; "NATO
did not keep its promise," he declared, that it would not expand
to the East, a promise that he said had been extended prior to
German unification.[10] Thus was demonstrated the importance of
a clearly defined NATO-Russia agreement.

Main Parameters

On May 24 and 25, 1994, upon the occasion of the DPC meeting
at NATO to which he had been invited (as a special instance of
privileged relations), Grachev proposed the "Main Parameters Of
the Practical Participation of Russia in the Partnership for Peace
Program." It was divided into three sections.

The first section, *Political Partnership*, accepted the invitation
to join PFP. "One of the goals of the partnership should be the
formation of a collective security and stability system in Europe.
The Partnership should contribute to the coordination of activities

of NATO and other structures in order to achieve the declared goals and not to the confrontation between them." These structures were "primarily" the OSCE but also the NACC, which should be transformed "into an independent [versus 'NATO-centric'] military and political cooperation body in the Euro-Atlantic area closely linked to the CSCE [OSCE]." A consultation mechanism should govern "the whole range of European and world security issues . . . both on regular and emergency basis." Russia would establish a mission at NATO headquarters and the Partnership Coordination Cell (PCC) at SHAPE. Cooperation would be carried out on peacekeeping, including "possible interaction during peacekeeping operations carried out by Russia in the former USSR" (an important statement in light of Russia's view of its "special role" in the CIS) and "participation in activities aimed at achieving interoperability of forces and financial resources provided by partners for peacekeeping operations."

In addition, Russia was ready to become a permanent member of the NATO bodies for counterproliferation in order to coordinate policy and undertake specific pilot projects in defense conversion. Grachev also noted that "Russia is interested in equal and non-discriminatory terms for the participation of Russian defense enterprises in the world trade in competitive and technology-intensive products designed for use, in particular, for peacekeeping, search and rescue and disaster relief operations."

The second section, *Military Partnership*, began with the "Coordination of Joint Military Activities." This would include consultations not only in the PCC but at the "Headquarters of the Allied Theater Forces in Europe" (the Major NATO Commands, which Poland had also requested). A permanent communication channel would link the Russian Ministry of Defense and the SACEUR (also a Polish idea albeit applied to Poland). Russia requested participation in the PMSC "to settle general problems of the partnership, ensure the required transparency of Individual Programs of Partnership, consider, when it is appropriate, the Partnership Work Program, study issues related to IPP, consolidate the IPPs and, if necessary, revise them, and to exchange information on military planning." (Was the Russian military seriously contemplating obtaining a *droit de regard* over other partner requests?)

The next subsection, "Peacekeeping Activities," included joint analysis of conflict or potential conflict situations, joint exercises both in Russia and in partner states with Russian participation, and interoperability of command and control systems of communication, reconnaissance, and air defense but "applicable only to the systems engaged in peacekeeping operation" (presumably meaning that by joining the PFP Russia would not undertake a wholesale conversion of its forces to NATO standards). The next subsection, "Transparency of Military Planning and Military Activities," foresaw Russian participation in the meetings of the NATO principal authorities on issues concerning arms control, military doctrine and strategy, force development, defense budgets, and exchange of military publications. The last subsection, "Training," called for consultations and exchanges. On May 24 Grachev indicated, in response to a direct question by Secretary of Defense Perry, that Russian peacekeeping operations could fall under NATO command.

Finally, *Other Areas of Cooperation* called for consultations on defense conversion and on humanitarian, search and rescue, and disaster relief. It also included consultations on a "joint unit for continuous monitoring . . . and providing timely response in case of emergencies" at NATO headquarters, development of a joint satellite system for environmental monitoring and the early detection of imminent disaster, and consultations on ecology.

At the same time, while offering this not insubstantial menu of cooperation, Grachev complained that NATO did "not conceal that its main task is to preserve the Alliance as the most powerful military structure on the continent." He repeated the line that OSCE and NACC should "coordinate" NATO, WEU, EU, and CIS efforts "in the domain of consolidation of cooperation and collective security." The PFP "is not a comprehensive response to the reality of the new epoch" but only "a first step toward forming a strategic community of allies as envisaged in the 'Main Parameters' paper." NATO had not yet agreed on producing a second statement outlining the NATO-Russia relationship, but on June 10, 1994, President Yeltsin called for a "protocol" defining the relationship that he implied NATO was not yet ready to conclude but whose adoption would be a precondition for Russia's signing on to the PFP.

Consultations "as Appropriate"

Meeting in Istanbul on June 9, 1994, the NAC formulated its approach as follows:

1. Despite its wish to strengthen relations with each of its partners, NATO retained its "right to take its own decisions on its own responsibility by consensus of its members, including decisions on the enlargement of the Alliance";

2. Russia was urged to develop "an extensive and far-reaching Individual Partnership Program, corresponding to its size, importance, and willingness to contribute to the pursuit of shared objectives." The allies were interested "in a broad dialogue with Russia in pursuit of common goals in areas where Russia has a unique or particularly important contribution to make"; and

3. Ukraine was also urged to be nuclear-free, and its adherence to the PFP was welcomed. Suggesting differentiation in a negative sense, however, UK foreign secretary Hurd linked alliance acceptance of the Ukrainian presentation document to a non-nuclear Ukraine (a status formalized in December 1994).

Meeting the next day in the NACC context, the delegates were unable to meet Russian insistence—dragged out over five hours—that the PFP would change NATO into a collective security organization and that changes were required to the 1990 CFE Treaty to allow for higher equipment holdings in the Leningrad and North Caucasus military zones. Russia, on the other hand, refused to agree to wording stating that the alliance was open in principle to new members, whereas France, Germany, and Italy seemed to be more concerned about language noting the WEU decision on associate partnership. Earlier the Russians had sought to borrow language from the NATO summit documents; for example, NATO was moving "to adapt itself to the new circumstances" to connote its proclaimed vision of the new security order. This prompted the Poles to declare they would also interpret NATO summit language, such as: "Active participation in the [PFP] will play an important role in the evolutionary process of the expansion of NATO."

In the end, although the NAC "welcomed" the WEU decision as complementary to the NACC (in fact the two were based on different philosophies, as previously discussed), the NACC, because of Russian objection (and Russia's not being a WEU partner), merely "took note" of the initiative. Nothing was stated in the NACC declaration about NATO expansion, whereas Polish foreign minister Andrzej Olechowski formally stated at the meeting that however important the NATO-Russia relationship, "it would be a paradox if the ties to be established between NATO and Russia were stronger and closer than those between NATO and the countries whose determination made this new relationship possible [by defeating Communism]."

According to key Polish participant Robert Pszczel, the experience raised the question not only whether the NACC would continue to issue statements but how frequently its meetings should take place. Others viewed the altercation as the exception and efforts to change the NACC as an overreaction. Nevertheless, toward the end of July 1994 discussions were already taking place about whether NACC declarations should be discontinued (a proposal that met with no Russian objections) and whether a fixed agenda of no more than two or three concrete topics should be set for each meeting. Specifically, on July 27, 1994, Canada and Hungary put forward ideas about how the form and content of NACC ministerials could be improved; they addressed the actual structure of meetings—the form of the agenda, the need for an NACC ministerial declaration, and the timing of meetings. In their view, the present format of prepared statements had proved to be largely insipid, and meetings were marked not by substantive interchange but by tedious discussions over communiqué language.

Despite this turbulence, Kozyrev finally signed the framework document on June 22, 1994, with Secretary Christopher attending. (Contradictions persisted in Russian policy: from June 20 to 23, 1994, another Russian-U.S. naval exercise, "Cooperation at Sea '94," was taking place south of Vladivostok.) No "protocol" outside the PFP, "not even in inverted commas," was signed. This was said to reflect Western concerns about the more formal nature of the word "protocol" in English. Instead, a "Summary of Conclusions of Discussions between the North Atlantic Council and Foreign Minister Andrei Kozyrev" was issued that acknowledged Russia's "weight and responsibility as a major European, international and nuclear power." It noted, as previously settled on June 15 in the NAC, agreement to share information, engage in political consultations "as appropriate," and cooperate in security-related areas "as appropriate" in the peacekeeping field. The Russian version of the text, however, used the more formal word "protocol." At the same time, NATO assured Moscow that it would abide by a policy dubbed "no vetoes, no surprises." Moreover, Christopher announced that "all the members of the PFP [now including Russia] are possible, prospective members"—which, as noted, London viewed as "inconceivable" for Moscow.

Following the signing ceremony at NATO, Kozyrev recalled President Yeltsin's December 1991 message expressing the desire for active cooperation with NATO and holding open the possibility of Russia's joining NATO at a certain stage. The opportunities of the current cooperative stage should first be exhausted, he said, "but there should be no haste in this. So far, there is a lot that can be done at this stage."[11]

6

How and Why

Reassurance

A moment of some drama occurred on the afternoon of Friday, June 17, 1994. The Poles had learned that U.S. secretary of state Christopher would be coming to Brussels the following week, solely to participate in the signing ceremony of Russian adherence to the PFP framework document. Poland was eager to balance this unique event—the presence of the U.S. secretary of state at a PFP signing—with a clear statement from President Clinton that Poland *would* become a member of NATO. Specifically, Poland wanted the president, during his Warsaw visit scheduled for July 6 and 7, to state that his administration viewed Polish membership in the alliance as "inevitable." If the Russians could have their "special" relationship, then the Poles should be given a green light for membership.

Poland also sought clear language in the IPP preamble (an innovation coming from the Polish side), which it hoped to have signed prior to Clinton's visit, that would at least acknowledge Poland's aspirations to join NATO or, even better, register a *mutual* NATO-Polish objective to realize this goal. NATO diplomats and officials offered conflicting advice: some counseled caution in creating a precedent; others urged Poland to "go for it." Familiar anxieties were resurfacing, with the Polish defense minister's complaining that despite growing cooperation with NATO, including "integration of communications systems, unification of procedures . . . tactical signs, maps, exercises with maps, computer exercises, the command language," and so forth, Poland could not at that point see the endpoint. "Up to now, no criteria have been defined that have to met by countries to be included in the NATO family."[1] He expressed concern that Western thinking was "very much dominated by inertia, especially in the context of our sense of a shortage of time. A game is in progress in which Russia is the first on the list to push the brake pedal." He cautioned that an enlargement process limited only to peacekeeping and humanitarian missions, one excluding measures that bring Poland closer to NATO structures, "will be a clear warning signal that we are taking part in a game without stakes.[2]

Even accounting for Central European anxiety, when the IPP was accepted on July 5, 1994, Acting Secretary General Balanzino did not even *mention* membership in his opening remarks, and the preamble recorded Polish commitment only to seeking full NATO membership. Finally, Poland also sought a substantial financial contribution for its PFP program; Polish officials had requested $25 million during an official visit to Washington in April 1994, which included lobbying Congress on NATO membership and assistance. Poland would match that contribution, or request a corresponding sum from Germany. Immediately following the April visit, Jeffrey Simon of the National Defense University put together, at the instruction of Joseph Kruzel, an interagency group that used as a base $200 million for PFP assistance for the forthcoming year, one fourth of which would be specifically allocated to Poland. This project was informally referred to as "priming the PFP pump," and Simon's Institute for National Strategic Studies cautioned in May 1994 that unless resources for the PFP were forthcoming the program could become "stillborn."[3]

As noted, until this point the U.S. position on when and why NATO would enlarge was unclear, obscured by such catchphrases as the "best possible future" or "hedging bets" on which way Russia would go. A clearer position, however, was signaled by the principal U.S. deputy assistant secretary of state for European and Canadian Affairs, Alexander Vershbow, on June 21, 1994, at a joint U.S. Atlantic Council–WEU Defense Ministers' meeting at the State Department, in which members of the NAA Political Committee participated. Said Vershbow: "Some people think NATO expansion more likely only if Russia moves away from reform. But there are other scenarios."[4]

Zbigniew Brzezinski described the new "more realistic U.S. policy" as seeking simultaneously to promote cooperation with

Russia and enhance Central European security rather than hold-ing the two as incompatible.[5] Indeed, it was said, President Clin-ton had actually "gone beyond his mandate" in his Prague state-ment just after the summit that NATO enlargement was not a question of "if" but "when" and "how"; thus the bureaucracy would have to play catch-up. Nevertheless, this evolution in U.S. policy would proceed in fits and starts, reflecting the lack of an agreed interagency position and the role of personalities.

Fresh initiatives emerged for helping to finance the PFP. On July 6 it was announced that Estonia, Latvia, and Lithuania would be entitled to a $10 million fund requested by the White House in the 1995 budget to support such regional peacekeeping efforts in Central and Eastern Europe as the combined Baltic peacekeeping battalion. Eligibility in the U.S. foreign military sales program would be included. The next day President Clinton announced his intention to seek, beginning in fiscal year 1996, $100 million in bilateral programs to support the PFP, with $25 million made available to Poland—this in addition to the Baltic fund. These resources could be used for "training, including English language training, participation in PFP exercises, equipping of designated units to improve their ability to function within the Partnership, and other activities in support of the Partnership's objectives."[6]

In an anxiously awaited speech before the Polish Sejm on July 7, President Clinton appeared to bury the "threat-driven" approach often resorted to by his officials and seemed to abandon the oblique "no new lines" rhetoric with preventing "a veil of indifference":

> [NATO] expansion will not depend upon the appearance of a new threat in Europe. . . . We are working with you in the [PFP] in part because the United States believes that when NATO does expand as it will, a democratic Poland will have placed itself among those ready and able to join. . . . we will not let the Iron Curtain be replaced with a veil of indifference.[7]

Earlier, on July 1, President Clinton had also suggested that dur-ing 1995 a "timetable" and "standards" for new NATO members should be developed—despite earlier NATO agreement at the Brussels summit that enlargement would depend on "political and security developments in the whole of Europe"—which obvi-ously could be interpreted as conditioned on events in Russia.

Puzzling Evidence

It was reported at the time of President Clinton's visit to Warsaw that "a conflict in Washington has been gathering momentum very clearly between the State Department on the one hand and the Senate, joined by a group of presidential advisors [citing National Security Council chief Anthony Lake and Perry] on the other." It was claimed that the State Department was concerned that early admission to NATO of the Visegrad countries would antagonize Russia and that legislation naming countries should include Russia or be withdrawn, whereas the other group was concerned about foreign policy's being dictated by Russian "phobias" that could strengthen Russian dominance in the region.[8]

Consequently, various congressional attempts to maintain enlargement momentum were defeated in conference "due to the lack of enough support in the House and vigorous lobbying on the side of the Administration," according to a Polish diplomat in early August 1994. This scuppered the "NATO Participation Act" moved by Senators Paul Simon and Hank Brown and adopted by the Senate on July 15, 1994, on a 74 to 22 vote; the act governed the transfer of excess defense articles to Poland, Hungary, and the Czech Republic.

Obviously, this fractured picture hardly impeded Russia from suggesting to NATO aspirants that they were only making fools of themselves, and that their best interests were to pursue good and cooperative relations with Russia. Outspoken Russian ambassador to Warsaw, Yuri Kashlev, suggested that "it is not we who tell Clinton or Kohl when to admit Poland. They simply act for their own reason, not Russia's. At the moment, they are not ready to admit Poland to NATO," and all that "rich America" could offer was an amount equivalent to "one aircraft" (actually, one-fifth the price of an F-15). He further argued that "there is no need to make NATO a leading or main organization, or the only one. That is why we are not in a hurry to join NATO and do not want other countries to do that."[9] Duma Defense Committee chairman Sergei Yushenkov believed that by now Central and Eastern Europeans had come to realize their "error" and that Russia was prepared to develop cooperation with them, including the supply of weapons.[10]

In September 1994 the Russia question arose once again. Speaking at the "New Traditions" conference in Berlin, Defense

Minister Rühe urged alliance expansion to the eastern frontier of Poland and a clear Western statement to Russia that NATO can cooperate with Russia but not integrate it in the alliance: "It would be wrong to pursue a policy that is determined by the highest-possible degree of ambiguity."[11] Secretary Perry refused, however, to rule out the possibility of Russia's joining, citing its nuclear arsenal as meaning that "all of our thinking with regard to Russia has to keep that fact front and center."[12] This struck some Europeans as indicating an attachment to the era of confrontation managed by nuclear arms control agreements. (Did the statement imply that at least some in the U.S. administration were experiencing sleepless nights over whether NATO expansion would cause Russia to abrogate nuclear arms control agreements or launch a preemptive strike against new NATO members?)

In any case, reportedly a "curious alliance" was forming against Rühe that included Pentagon officials, who were wary of new security obligations and of NATO expansion as playing to President Yeltsin's opponents, and the German Foreign Office, which preferred EU expansion.[13] In July 1994 deputy assistant secretary of defense Kruzel predicted that "there will be new members of the Alliance, some earlier, some later," but he did not foresee Russia's becoming a member. Such a large state, he believed, would not subordinate itself to an alliance led by the United States. For the same reason, Kruzel did not envisage Russia's submitting its defense planning for review at NATO. The PFP for Russia would probably serve instead as a "way for them to be reassured that NATO is not plotting against them."

Nevertheless, pro-enlargement forces had by no means been defeated. Despite earlier setbacks, on October 8, 1994, the Senate gave final passage to the "NATO Participation Act." Although it had not survived as an amendment to the Foreign Operations Bill or to the Defense Appropriations Bill, it did pass as an amendment to the "International Narcotics Reauthorization Bill" (an anti-drug abuse act). According to Senator Brown: "This amendment reverses the Clinton administration's policy of excluding Poland, Hungary, the Czech Republic and Slovakia from NATO membership. It is a signal to the world that the United States wants to move toward NATO membership for those countries." The bill became law on November 4, 1994.

The act first specified, as the sense of the Congress, that

active PFP partners "should be invited to become full NATO members in accordance with Article 10 of [the North Atlantic] Treaty at an early date" if they "maintain progress toward establishing democratic institutions, free market economies, civilian control of the armed forces, and the rule of law" and "remain committed to protecting the rights of all their citizens and respecting the territorial integrity of their neighbors." The act authorized the president to assist the four nations "and other Partnership for Peace countries emerging from communist domination" to facilitate the transition by supporting, among other things, joint training, greater interoperability, conformity of military doctrine, and the transfer of both lethal and nonlethal surplus U.S. military equipment.

An unnamed State Department official reacted, "We didn't want to differentiate and set a timetable for membership in NATO for any country in Eastern Europe."[14] As noted, however, fewer than four months earlier, President Clinton while in Warsaw had seemed to indicate just the opposite.

Momentum increased, given that the foreign policy plank of the *Contract with America*, the Republican congressional platform, supported NATO expansion to the Visegrad no later than January 1999. The Republicans assumed leadership of both houses of Congress when the 104th Congress convened in January 1995.

On September 9, 1994, Alexander Vershbow, then senior director for European Affairs at the National Security Council (and appointed ambassador to NATO three years later in September 1997), stated that "the time has come for the NATO countries to begin to discuss among themselves the criteria and the timeline that may be associated with the expansion of the Alliance." On October 27, 1994, "it was no accident" that an article appearing in the *New York Times* cited a senior administration official (rumored to be new assistant secretary of state for European and Canadian Affairs Richard Holbrooke, who achieved prominence a year later by having compelled the warring parties in former Yugoslavia to arrive at a peace settlement) as conceding: "Much of the rhetoric that was used in the past was empty of substance. Partnership for Peace is like getting guest privileges at the club—you can play golf once in a while. Now we want to send the bylaws and ask, 'Do you want to pay the dues?' "

According to a senior congressional staff member that same month, although the "when" issue was "inextricably linked to 'how,' " four reasons could be cited for the perceptible shift in administration tendencies:

- The Clinton administration members may have felt vulnerable on the issue, given the Republican foreign policy challenge, thus deciding to work together;
- As already noted, when President Clinton stated in Prague that the issue was not whether but how and when NATO would expand, he went beyond agreed U.S. policy, making follow-up inevitable;
- Personalities had changed in the administration from "cautious" officials to "movers and shakers," including many "closet NATO expansionists"; and
- "Just maybe" the administration recognized that criticism was merited, with the new assistant secretary Holbrooke having praised elements of Senator Lugar's June 28, 1994, address, in which the senator termed PFP a "policy for postponement."

By early November it was reported that the Clinton administration, while still delaying the "who" and "when" issues, was developing "precepts" for NATO membership in a "near-term goal . . . to get the Alliance to agree to begin a formal process, aimed at defining what it will take to expand."[15] At the NAA Annual Session in Washington, November 14–18, 1994, the Political Committee was exposed to a snapshot of the debate from Zbigniew Brzezinski and Joseph Kruzel. (On August 19, 1995, Kruzel was one of three U.S. officials who died in an accident on the Mount Igman road in Bosnia.)

Brzezinski argued that the criteria for membership should be primarily "geopolitical"—for example, commitment to democracy and shared values, an absence of ethnic or territorial disputes with neighbors, territorial contiguity with other NATO members—and not technical. Technical aspects could be resolved *after* membership. He cautioned:

> I suspect that the inclination of some people, including within the United States, to emphasize primarily operational, technical criteria is a reflection of a hidden but capricious intention to delay the expansion of the Alliance. . . . there is a deliberate effort by those who have lost the struggle on the principle itself to refight the struggle on the level of qualifica-

tions, by raising these qualifications to degrees which cannot be met in the near future.[16]

The former U.S. national security adviser also counseled that the question of Russian membership in NATO need not be addressed at the present time. Whether Russia would one day be able to work with the NATO allies on the basis of consensus was a question for the future. The expansion of NATO and a treaty of cooperation between Russia and NATO should be pursued together with two emphases: NATO expansion is not an anti-Russia program "but a logical consequence of the growth of Europe," and expansion is "not designed to exclude Russia but to engage Russia in a larger system of cooperation and security."[17]

Kruzel was, however, more cautious and stressed the collective defense dimension:

> We cannot give you a checklist that you can tick off, and when you completed your last requirement you are admitted. But there are two things that I would insist on of any prospective member of the Alliance. The first is to show us you are worth defending. Show us that you share our values. Second, show us that you bring something to the table, that you can make a contribution to collective defence. Otherwise, Article Five of the North Atlantic Treaty, in the context of declining defense budgets but expanding commitments, could be made hollow.
>
> We in NATO have a lot of work to do before we are ready for expansion. We need to think about how to bring these countries into the military command structure. We need to think about forward-defense—will we need to station forces in the territory of these new members, will we need to put nuclear weapons there, preposition equipment there? How will we take on this challenge of defending a considerably increased strategic space with a considerably reduced defense posture?[18]

The NAA itself endorsed, albeit narrowly, on November 18, 1994, a resolution moved by U.S. senator William Roth, Jr. (Republican-Delaware), stating that "several CEE [Central and Eastern European] countries have made major progress towards satisfying all reasonable requirements for membership in NATO" and that "the Alliance should establish the terms of reference, criteria, and timetable for the early admittance of these countries."[19] Although no countries were named, the NAA urged that memoranda of understanding elaborating the protocol of accession should be reached no later than mid-1995 and that the membership process should commence within the next two to five years.

Of course, even assuming that the Clinton administration was now firm on NATO enlargement, there were the 15 other allies to consider. On November 21, 1994, French prime minister Edouard Balladur observed: "The aim is not to speed up the enlargement of security bodies such as NATO or the WEU. Everyone is well aware that the sudden inclusion of new countries in these alliances could cause more instability than stability on our continent."[20] It was also reported that the EU had suggested to the Baltic states that they adopt neutrality.[21] A leaked cable at the end of November from German NATO ambassador Hermann von Richthofen was remarkable as the first disclosed official communication to confirm an American "Russia-first" policy:

> Unilateral changes in U.S. policies towards the ex-Soviet Union and eastern Europe since the NATO Summit in January directly affect the Alliance. There was an abandonment at short notice, without any consultation, of the "Russian First" concept . . . if the U.S. on the one hand promotes a robust continuation of Partnership for Peace, but if on the other hand the concept is described by parts of the administration as "inadequate" and "oversold" after a few months, then this part of alliance policy ends up being discredited. . . . Now the U.S. is applying political pressure for more speed in the expansion question, without paying attention to how this affects internal alliance positions on eastern policies as well as the military and financial consequences.[22]

A Process of Examination

On December 1, 1994, the NATO foreign ministers in Brussels arrived at the following decision:

> We have decided to initiate a process of examination inside the Alliance to determine how NATO will enlarge, the principles to guide this process and the implications of membership. To that end, we have directed the Council in Permanent

Session, with the advice of the Military Authorities, to begin an extensive study. This will include an examination of how the Partnership for Peace can contribute concretely to this process. We will present the results of our deliberations to interested Partners prior to our next meeting in Brussels [one year later]. We will discuss the progress made at our Spring meeting in the Netherlands [although the United States had sought the Spring date as the deadline for the review].

However, the alliance explicitly stated that it would *not* examine the "who" and "when" questions: "We agreed that it is premature to discuss the timeframe for enlargement or which particular countries would be invited to join the Alliance." All new members "will be full members of the Alliance," and enlargement would be decided on a case-by-case basis and "some nations may attain membership before others." However, "we have no desire to see the emergence of new lines of partition," and a cooperative European security architecture required "the active participation of Russia." Among partners Russia was singularly offered a meeting with the council during the biannual regular ministerial meetings "whenever useful."

In short, *nothing remotely specific was decided.* A Polish diplomat concluded, perhaps overdramatically, that the Central European countries were being "pushed into the dustbin. We are being cheated. There is no balance. This is very distressing." In the NACC meeting on the following day, Foreign Minister Olechowski welcomed "wholeheartedly . . . the decision to set the enlargement process in motion," but cautioned against delay and "any attempt to base the 'European security architecture' again on the balance of power principle."

On December 1 Foreign Minister Kozyrev was supposed to conclude the Russian IPP and a program of consultations and cooperation. After a phone call to President Yeltsin, however, he refused to conclude either, claiming he had been "surprised" by the decision to study enlargement (even though the Russians had been informed of the text in advance). Kozyrev's behavior was explained by NATO officials and diplomats as probably designed for domestic consumption, as was President Yeltsin's presentation on December 5 at the CSCE summit in Budapest (although Yeltsin used arguments similar to those NATO and NATO national authorities had used regarding "best possible futures"):

"Europe, even before it has managed to shrug off the legacy of the Cold War, is risking encumbering itself with a cold peace." NATO's search for a new role should not create new dividing lines to which, he said, NATO expansion ran counter. "Why sow the seeds of mistrust? After all, we are no longer enemies, we are effectively all partners."

On the same day at the Budapest summit, however, President Clinton affirmed that the alliance had started work on the requirements for membership and that NATO would "not automatically exclude any nation from joining." At the same time, no country outside would be allowed to "veto" expansion, a process that, he continued, would make NATO's new members, old members, and nonmembers more secure.

Nevertheless, in Brussels Kozyrev stated that he was prepared to return to a discussion on broad partnership and did not object to "increasing the number of members of the alliance," provided this followed the "important stage" of partnership with Russia (whereas in Budapest Yeltsin even mentioned possible Russian NATO "political membership").[23]

The December 20 statement of Russian Foreign Ministry spokesman Grigori Karasin may therefore have been *nye sluchaino* (no accident). Moscow was satisfied, he said, with the "well-thought-out and balanced" NAC communiqué, welcoming its emphasis on "Russia's potential for making a significant contribution to ensuring stability and security in Europe" and its view of NATO enlargement as part of the "development of larger cooperation for security in the Euro-Atlantic region as a whole," which must "contribute to common security."[24] But the difficult NATO-Russia relationship was by no means resolved.

As 1995 began, President Clinton reflected on the debate, candidly conceding on January 13 that sentiment for shutting out Central Europe from the West was hardly quiescent:

> Some argue that open government and free markets can't take root in some countries, that there are boundaries—that there will necessarily be boundaries to democracy in Europe. They would act now in anticipation of those boundaries by creating an artificial division of the new continent. Others claim that we simply must not extend the West's institution of security and prosperity at all, that to do so would upset a delicate balance of power. They would confine the newly

free peoples of Central Europe to a zone of insecurity and, therefore, of instability.

I believe that both those visions for Europe are too narrow, too skeptical, perhaps even too cynical. . . . Countries with repressive political systems, countries with designs on their neighbors, countries with militaries unchecked by civilian control or with closed economic systems need not apply. . . . But NATO expansion should not be seen as replacing one division of Europe with another one. It should, it can, and I am determined that it will increase security for all European states, members and non-members alike. In parallel with expansion, NATO must develop close and strong ties with Russia. The Alliance's relationship with Russia should become more direct, more open, more ambitious, more frank.[25]

7

From Noordwijk to Brussels

On May 30, 1995, the NAC convened in Noordwijk, the Netherlands, to review the first results of the "philosophical" NATO internal study on the "how" and "why" of enlargement. The ministers repeated the open-ended formula that the objective of NATO enlargement would be to "enhance security for all countries in Europe, without creating dividing lines." They also noted "steady, measured progress" on the internal report—which for Russia meant "hasty." The report, which would be completed by September 1995, would be briefed to interested partners, and their reactions would be reviewed (although in fact the briefing teams, under a limited mandate, asked few questions). At the next ministerial in early December 1995, NATO would "consult on the way forward in our examination of enlargement."

The allies disagreed, however, even on when the briefings to partners should begin, with the United States favoring an earlier start. France and Spain differed with the allies on the meaning of the agreed language that "all new members of NATO will be full members of the Alliance" (but not without good reason: if new members would have to be integrated into the military command, it would constitute a double standard).

Although some in Congress and elsewhere increasingly began to debate NATO enlargement in terms of preserving the sanctity of a "contract to go to war" including nuclear response to aggression, the NATO Participation Act Amendments of 1995, first tabled in the Senate on March 23, 1995, took both the military defense and democratic integration aspects into account:

• NATO remained "the only multilateral security organization capable of conducting effective military operations to protect
Western security interests";
    • Threats to the new democracies "would pose a security
threat to the United States and its European allies"; and
    • NATO expansion "can create the stable environment
needed to successfully complete the political and economic transformation envisioned by European states emerging from communist domination."

Indeed, when the Senate considered the Washington Treaty
in 1949, the Foreign Relations Committee stated: "[The alliance]
will free the minds of men in many nations from a haunting
sense of insecurity and enable them to work and plan with that
confidence in the future which is essential to economic recovery
and progress."[1] Likewise, during the height of the Cold War, a
1956 alliance study noted, in words extremely relevant today in
light of questions that continue to be raised about the "why" of
NATO and its enlargement:

> While fear may have been the main urge for the creation of
> NATO, there was also the realization—conscious or instinc-
> tive—that in a shrinking nuclear world it was wise and timely
> to bring about a closer association of kindred Atlantic and
> Western European nations for other than defense purposes
> alone; that a partial pooling of sovereignty for mutual protec-
> tion should also promote progress and cooperation generally
> [toward] the development of *an Atlantic Community whose roots
> are deeper even than the necessity for common defense.* This implies
> nothing less than *the permanent association of the free Atlantic
> peoples* for the promotion of their greater unity and the protec-
> tion and the advanceent of the interests which, as free de-
> mocracies, they have in common.[2]

In dramatic political contrast to the noncommittal NATO approach, earlier that year on February 16, 1995, the U.S. House of
Representatives passed H.R. 7, the bill including the "NATO
Expansion Act," by a vote of 281 to 141. This bill, which I cosponsored, was one of the first brought forward for consideration by
the new Republican majority in the House.
    H.R. 7 stated that admission to NATO of Central and Eastern
European countries recently freed from communist domination
could contribute to international peace and enhance the security

of those countries. Poland, Hungary, the Czech Republic, and Slovakia should be invited to become full NATO members, and alliance governments and NATO should furnish appropriate assistance to facilitate the transition. The bill also provided that any other European country emerging from communist domination could, upon meeting Article 10 requirements, be invited to join. These countries (defined as Albania, the Baltic states, Bulgaria, Romania, and "certain countries" that were part of the former USSR and Federal Republic of Yugoslavia) could become members if they

• met the criteria of "shared values and interests," democratic governments, free market economies, civilian control of the military, police, intelligence, and other security services;
• adhered to OSCE commitments;
• were committed to furthering the principles of NATO and contributing to the security of the North Atlantic area and accepted the "obligations, responsibilities, and costs" of NATO membership; and
• remained committed to protecting the rights of all their citizens and respecting the territorial integrity of their neighbors.

The Republican platform in the 1994 congressional campaign, *Contract with America*, had contained a date certain—January 10, 1999—for the entry of Hungary, Poland, the Czech Republic, and Slovakia into full membership in NATO. The original proposal also had not gone any further than mentioning the four Visegrad countries as possible candidates. But although there was indeed some variance between the original proposal and the language adopted, the movement was clearly in the same direction, despite efforts to make the assistance program merely discretionary, to delete mention of the Visegrad states, and to argue that Central Europe was irrelevant to U.S. national security.[3] The most important thing then was simply the fact that one house of Congress had gone on record as endorsing NATO enlargement. The Senate completed hearings on the bill on March 21, 1995.

The NATO Participation Act Amendments of 1995 in the Senate identified those countries "most ready for closer ties with NATO," such as Poland, Hungary, the Czech Republic, and Slovakia, but also stated that NATO membership could eventually include Albania, the Baltic states, Ukraine, Romania, Bulgaria, and Slovenia. It also called upon the U.S. government to urge *observer status in the NAC* for countries designated as eligible for NATO transition assistance—an idea close to what the NAA had proposed in 1990—although this provision, opposed by the administration, was dropped by the summer.

In 1995 PFP exercises increased from 3 in 1994 to 11 (including the first in the United States—*Cooperative Nugget* in Louisiana and *Cooperative Support* in Norfolk), and 14 countries were now participating in a defense planning and review process (PARP) that is reviewed every two years; NATO provided 20 "interoperability objectives" for the first two-year cycle for the forces associated with PFP activities—an important confidence-building measure for transparency modeled on NATO's own defense review process (Russia did not join, despite the "Main Parameters" proposal for exchanges on defense planning—which partly already existed in OSCE documents). Bilateral activities included the 1992 U.S.-funded IFF program, *Peace Pannon*, which was turned over to Hungary in December 1994 and included the installation of transponders on MiG-21, MIG-23, and MIG-29 aircraft; Bulgaria's authorizing of the firing of surface-to-air missiles by Greek air defense units near Varna; and continued progress in the U.S. Regional Airspace Initiative, designed to bring Central and Eastern Europe in line with NATO operational standards.

Thus, even as NATO leaders hesitated on new alliance membership, events on the ground and in the air were building the infrastructure for what Jamie Shea termed "bottom-up" integration.

All the same, in Noordwijk Polish foreign minister Wladyslaw Bartoszewski urged yet again the need for *differentiation*, and also drew attention to what he understood as a "lack of transparency surrounding the present debate on the pace of enlargement of the Alliance." This longstanding differentiation problem reflected earlier concerns elaborated by Polish diplomat Robert Pszczel during the summer of 1994: "The Allies instinctively gravitate towards unification of cooperation programs. . . . The temptation to fall back on the 'lowest common denominator, but including all Partners' is just too great." He argued, instead, that NATO should commence an "associated participation" method of work whereby Polish experts would be invited to "at least the margins" of the *regular* NATO committee meetings and Polish officers temporarily assigned to NATO *regional* commands.[4] Poland was making rapid progress in integrating with NATO in areas ranging from the conversion of common scale maps (from a scale of 1 : 250,000 to 1 : 1 million) to the adapting of command systems and the implementing of defense resources management, but much of this success was due to *bilateral* cooperation with nations such as Denmark, Germany, and the United States. In contrast, it was not until June 1995 that the second Polish IPP, submitted in November 1994 and with five times the number of activities as the first, was accepted by NATO; *two years later*, Polish deputy Longin Pastusiak was still compelled to observe that "relationships with some of our Western partners [i.e., the United States and Germany] have already acquired the quality of a quasi-alliance, and are in many cases of a more profound nature than with the Alliance itself."[5]

Hans Jochen Peters informed the NAA in Budapest on May 28, 1995, that distinguishing among partners, in his personal view, could cause problems: "If we establish different classes of PFP membership the result could be to weaken and not strengthen this important instrument." What, I was compelled to ask him, had then become of "self-selection"?

The United States also urged, as could be expected, greater dynamism. General John J. Sheehan, Supreme Allied Commander, Atlantic (SACLANT), informed the NAA on May 27 that for NATO to retain "its central role" and grasp "the historic chance it now has . . . of gradually becoming the principal guarantor of security, peace and democratic development in Europe," it must "embrace the PFP with tangible initiatives and exercises." Change toward that end, he stated, "will not happen if NATO's future is left to bureaucracies tied to the legacies of the past." For example, Romanian foreign minister Teodor Melanescu observed on May 30 that political dialogue in the NACC was losing interest for partners (although NATO communiqués continued to insist NACC would play "an integral part of the new European security architecture").

In Noordwijk Secretary Christopher also urged the NACC to go further, consistent with U.S. interventions over the last few years promoting additional movement shortly after alliance decisions had been taken. The United States recommended that the NACC

• establish a set of principles for civilian and democratic control of the military (inherently difficult on account of differing models within NATO countries themselves);

- extend joint defense planning and review to all armed forces of partners, not just to those dedicated to PFP purposes;
- suggest measures for adapting partners' military doctrine and forces to NATO's;
- find ways to enable partners to play "a more active, substantive role in the planning of Partnership activities and exercises";
- engage partners "more routinely in the substantive activities of the NAC and NATO senior committees"; and
- increase significantly those NATO resources dedicated to the PFP.

Cold Peace?

The Russian attitude remained problematic. Toward the end of 1994, there were indications that the Russian Foreign Ministry was prepared to concede to an enlarged NATO membership, at least for the Visegrad countries. In exchange, there would be territorial guarantees (concern was expressed about the Kaliningrad *oblast*) and a prohibition on the stationing of nuclear weapons or foreign forces—issues reportedly discussed during talks in late February between Russian deputy foreign minister Georgi Mamedov and U.S. undersecretary of state Strobe Talbott.[6] However, President Yeltsin demanded no compromises.

Among Central Europeans there were concerns that the prospects of enlargement were growing dimmer and that NATO had agreed to "freeze" even discussion of enlargement until after the December 1995 parliamentary and June 1996 presidential elections (if not for all of that year or beyond). There were also concerns that Russia would exploit the written pledge, reportedly made in March that year by President Clinton to President Yeltsin, that Russia would not be automatically excluded from NATO membership (President Clinton said the same at the Budapest summit):[7] "We are getting funny signals of a shift in the U.S. position," a Central European diplomat observed, "that the NATO-Russia dialogue on security may be molded in some way to encompass discussion of NATO's enlargement plans. Are we the hostages?"[8]

"Public opinion" and the forthcoming parliamentary (December 1995) and presidential (June 1996) elections in Russia were routinely cited by Russian officials and politicians as reasons for NATO "not to wave a red flag," as the vice chairman of the

Russian Duma CIS Committee, Deputy Yuri Voevoda, put it on June 15, 1995, during a visit to NAA headquarters in Brussels. Yet, a public opinion poll of Moscow and its regions found that fewer than half of respondents—44 percent—were "negative" toward the prospect of Eastern Europe in NATO, whereas 25 percent were "positive" and 24 percent "neutral."[9] (Voevoda dismissed the findings as reflecting the inward looking nature of those "inside the Moscow ring.") All the same, General Aleksandr Lebed, former commander of the Russian 14th Army in Moldova and now leader of the *Congress of Russian Communities* and popular presidential candidate, concurrently predicted that the enlargement of NATO would unleash "World War III" but that Russia had a "military secret" to prevent expansion and NATO's alleged designs on all of Russia: "But they will not succeed."[10] Such was the level of discourse emanating from some promoters of great power assertiveness, *derzhavnost*, in the Russia that NATO sought to engage in "strategic partnership."

Apart from Russian obstructionism, some European allies encouraged a slowing down of the enlargement process. Italian foreign minister Agnelli stated on April 12 that "it is necessary to approach this problem [NATO enlargement] gradually and to allow time for reflection,"[11] a course (not surprisingly) endorsed by Kozyrev. On May 16, 1995, the Portugese foreign minister José Manuel Durao Barroso stated that "it is important that this expansion should not mean any threat for Russia"[12]—as if there were reasonable doubt that enlargement could pose such a threat. It was also reported that the "how" and "why" study could be prolonged into 1996, for one reason because "consultations with the aspirant nations could unearth differing interpretations of NATO responsibilities that, unless resolved, could divide alliance members themselves"(for example, what is a "full member?").[13]

On May 10, 1995, at the Clinton-Yeltsin summit meeting in Moscow, President Yeltsin noted that an exchange of views "on the NATO problem" had taken place but "we have found no solution." Yeltsin suggested, again, that the issue not be rushed and that consultations continue in mid-June 1995 at the G-7 meeting in Halifax and in October at the fiftieth anniversary of the United Nations in New York, where "perhaps a final accord will emerge." Clinton stated that it was not important that Russia and the United States come to an agreement right away and noted that "even NATO has not made such a decision . . . who knows,

perhaps [sic] there are still disagreements within NATO itself"[14]—
an admission of reality, if not exactly a clarion call of leadership.

Nevertheless, a small breakthrough did occur in Noordwijk
when on May 31, 1995, Kozyrev formally accepted both the IPP
and a special "broad, enhanced dialogue" with NATO. He
warned, "A decision about the enlargement of NATO to the East
would create for Russia the need for a corresponding correction"
of the Russian attitude to PFP,[15] but he also claimed that "the
question of NATO expansion to the East becomes no longer as
acute and political and is no longer on the agenda," that it would
be a routine issue to be discussed among many others at expert
level.[16] He suggested that NATO "halt and think rather than act
hastily and blindly."

On June 22, 1995, another appeal for slow motion came from
Senator Sam Nunn in a speech said to have stimulated second
thoughts among legislators:

> NATO was established primarily to protect the Western de-
> mocracies from an expansionist Soviet Union . . . In the
> longer term, we cannot dismiss the possibility of a resurgent
> and threatening Russia [which] is a vast reservoir of weap-
> onry, weapons material and weapons know-how . . . our
> first principle should be do no harm.
>
> NATO's announced position is that the question of en-
> largement is not whether, but when and how . . . [but] some-
> body had better be able to explain to the American people
> why, or at least, why now. . . . Are we really going to be able
> to convince the East Europeans that we are protecting them
> from their historical threats, while we convince the Russians
> that NATO's enlargement has nothing to do with Russia as a
> potential military power?
>
> . . . because a conventional military response from Rus-
> sia in answer to NATO enlargement is infeasible, a nuclear
> response, in the form of a higher alert status for Russia's
> remaining strategic nuclear weapons and conceivably re-
> newed deployment of tactical nuclear weapons is more likely.
> The security of NATO, Russia's neighbors, and the countries
> of Eastern Europe will not be enhanced if the Russian military
> finger moves closer to the nuclear trigger.

Therefore, Senator Nunn suggested a two-track approach toward NATO enlargement, with considerations about Russia taking front and center in both cases:

The first track would be evolutionary and would depend on political and economic developments within the European countries who aspire to full NATO membership. When a country becomes eligible for European Union membership, it will also be eligible to join the Western European Union and then be prepared for NATO membership, subject of course to NATO approval. This is a natural process connecting economic and security interests. We can honestly say to Russia that this process is not aimed at you.

The second track would be threat-based. An accelerated, and if necessary immediate, expansion of NATO would depend on Russian behavior. . . . We would be enlarging NATO based on a real threat. We would not, however, be helping to create the very threat we are trying to guard against. . . . If future developments require the containment of Russia, it should be real containment, based on real threats.[17]

In response, the leader of the Norwegian Conservative Party (*Høyre*) and NAA Political Committee General rapporteur Jan Petersen, whose nation is the only NATO country to share a border with Russia, urged Senator Nunn to reconsider and lend his prestige to a principled enlargement independent of what occurs in Russia:

First, NATO was not established, as [Senator Nunn] argued, "primarily to protect the Western democracies from an expansionist Soviet Union." This is a militarized view of NATO that encourages threat-based thinking and overemphasis on the implications of the security "guarantee" and *angst* about "hollow commitments"—as if the principle issue was whether Bratislava could be defended. To the contrary, "the Alliance evolved as a hybrid organization: one that maintained collective security against the Soviet Union and actively pursued the construction of a trans-Atlantic community of nations . . . that included a peaceful Germany. . . . This belief [of community] preceded the American drive to balance against the Soviet threat." Recall, after all, that SHAPE did not become operational until 1951.

Second, it is interesting that an American legislator would suggest prior EU membership for membership in the Atlantic Alliance. This could prove to be how events come about but cannot be a precondition. Although Senator Nunn argues that this connection would mean that we could "honestly say to Russia that this process is not aimed at you," this wrongly suggests an offensive nature to NATO requiring proof to the contrary. Moreover, it does nothing to change the Russian arguments that NATO should not enlarge under any circumstances.

Third, to say that we will enlarge at an accelerated pace if Russia becomes a threat lacks all credibility. If we are tenuous in our position during healthy relations, why would we be bold in crisis?

Instead, Petersen urged that 1996 inaugurate a *pre-accession phase* for Poland, the Czech Republic, and Hungary, followed by others including the Baltic states entailing representation at SHAPE and the Major Subordinate Commands, participation in NATO political and military meetings "as appropriate," preferential funding and participation in other NATO activities such as the NATO School in Oberammergau, and close working relations with the NATO Standardization Office. "This would send the missing political signal that enlargement can in the end acquire transparency and dynamism. Decisions on the future of PFP, NACC, the NATO-Russia relationship could be settled later and on their own merits."[18]

A sympathetic view was also taken by the House of Commons Select Committee on Defense in July 1995. The committee deemed NATO enlargement as "both necessary and desirable" and regarded the alternative of a security vacuum as "in the long run unacceptable." Arguments about not drawing new lines "simply fossilises the present line," and it was recalled that NATO had always had borders with the Soviet Union in Norway and Turkey. However, the MPs saw no reason for urgency: enlargement could, in their view, occur within the next 10 years. The committee also stated that "the rationale for extension is political; extension is a means toward an end, the end being a more stable European security environment," but that enlargement should not overshadow the needed reform of NATO structures to adapt to new risks and the likelihood that future operations would entail coalitions of the willing, including non-NATO, states.[19]

An influential Russian policy group, the Council on Foreign and Defense Policy, suggested that NATO enlargement was not urgent and, moreover, preventable altogether. NATO should continue to serve as "a guarantor of stability in the relations along the West-West axis" and could become a pillar of a new collective security structure. NATO as "a defensive military and political union . . . is not a military threat for a democratic Russia." Enlargement, however, would have negative political and psychological effects upon Russian domestic politics, risking turning Russia into a revisionist power. A decision on enlargement, if it could not be avoided, should be delayed by four years if a NATO-Russia treaty was to be concluded, and Russia should work with countries presumed opposed to enlargement—citing Portugal, Spain, possibly Italy, Great Britain, France, and Greece—and take advantage of what was described as "something close to a consensus in the leadership circles of NATO with respect to nonacceptance of NATO expansion." Russia should work to create an OSCE directorate comprising the relevant members of the UN Security Council, Germany, and representatives of EU, WEU, NATO, and the CIS, as well as pursue other versions of creating a "system of collective European security—for instance, on the basis of a special agreement or on the basis of transforming NATO into an all-European system of collective security with inarguable priority given to the inclusion of Russia."[20] However, although no NATO nation with the exception of France had articulated a similar "all-European security" concept (in the months preceding the July 1992 Helsinki OSCE summit), NATO had already in 1992 offered to make its resources available to support OSCE peacekeeping and implementation of UNSC resolution and by 1995 had proved to be the only organization able to end the violence in the former Yugoslavia.

In September 1995, at a NATO conference in Knokke-Heist on the Belgian seacoast in Flanders, Sergei Rogov took a stand both more critical and flexible. Rogov, the director of the Institute of USA and Canada Studies in the Russian Academy of Sciences, saw a NATO strategy to isolate Russia while expanding Western influence, even though the main issue for European security was not NATO enlargement but the NATO-Russia relationship. He described that relationship as "declaratory" rather than "real." He admitted, however, that "Moscow has been unable to specify what it expects from NATO," but it was not too late to establish a Russia-NATO "military-political partnership" on a legally binding basis for joint decision-making and cooperative implementation. In the case of NATO expansion, he called for the nonstationing of nuclear bases or foreign forces on the territory of new NATO members as well as a ban on NATO activities in the Baltic states and former Soviet republics. He also called for a NATO guarantee on the inviolability of Russian frontiers including Kaliningrad, for an updating of the CFE treaty, and for the developing of NATO-CIS relations.[21] In response, NATO assistant secretary von Moltke slightly veered from his normal composure by pointing to the various initiatives he had taken without meeting Russian response—and Kozyrev conceded subsequently that the "old guard" in Moscow had blocked any further discussion of compromise.

The Enlargement Study

On September 20, 1995, the North Atlantic Council endorsed the internal study, and briefings to partners were set for September 28. Concurrently, and before the briefing, NATO offered CFE Treaty concessions to Russia, the Council of Europe Parliamentary Assembly lifted its freeze on the application of Russia (brought on by Chechnya); a NATO-Ukraine "enhanced relationship" was announced; thought was being devoted on how to integrate Russian forces in a NATO-led monitoring of a peace plan in Bosnia of up to 60,000 troops (why Rogov spoke of isolation, then, is mysterious); and NATO offered Russia proposals for still further documents: a "political framework" and a work plan that NATO hoped would be ready by the end of the year. Nevertheless, Foreign Minister Kozyrev declared that Russia still opposed NATO enlargement and any "horse trading."

The 28-page "Study on NATO Enlargement" was naturally deliberately cautious. Nevertheless, it was welcomed by Central and Eastern European diplomats as raising frankly and honestly "all the right questions" and providing referencing: "If Russia wants to criticize enlargement, it cannot just continue to interpret summit language. The study is about the *enlargement* process," a Czech diplomat emphasized.

The 26 partners were addressed at NATO by Secretary General Claes, followed by NATO assistant secretary von Moltke, the assistant secretary general for defense planning and policy Anthony Cragg, and the director of the Military Committee, Lt. General Jan Folmer. During the whole affair, which took only one and a half hours, the Russian delegates did not speak once. These were the essential points:

1. *Why NATO Will Enlarge.* "To build an improved security architecture . . . without creating dividing lines"; to promote democracy, habits of cooperation and consensus building, good-neighbor relations, defense planning and budget transparency; to contribute to security; and to strengthen and broaden the transatlantic relationship.

2. *Principles of Enlargement.* All new members would enjoy equal rights and duties, settle ethnic or territorial disputes, contribute militarily to NATO missions, and work for consensus on issues of admitting new members. There would be no *a priori* requirement for stationing allied troops or nuclear forces on the territory of new members, but it would be important that allied forces could be deployed, when and if appropriate, on the territory of new members. New NATO headquarters on the territory of new members "may be required," and new allies should participate in "an appropriate way" in the NATO command structure.

3. *How NATO Will Enlarge.* Active PFP participation would contribute to preparing for membership. New members would not be required to join the integrated military structure, but would need to meet certain minimum standards that could be worked out through PARP. "The premature development of measures" to facilitate enlargement was cautioned against, but "there will come a point, after a country has been invited to join the Alliance, when specific measures for preparing the accession of that country will have to be devised." No link was made to EU membership or to a strategic partnership with Russia, but the study stated that "an eventual broad congruence of European membership in NATO, EU and WEU would have positive effects on European security," whereas "development of the NATO-Russia relationship, and its possible formalization, should take place in rough parallel with NATO's own enlargement"—language that provided the only hint of how the "who" question could be answered and thus actually inched somewhat beyond the "how" and "why" mandate. Although there had been discussion about whether to rule out Russian membership, the study declared, in a paraphrase of the North Atlantic Treaty, that "any European state" could apply. NATO teams then set off to capitals to present further briefings on the study, but their mandate in responding to questions was limited.

Willem van Eekelen cautioned that "the worst thing to happen would be a rejection of ratification by one or more of the parliaments of NATO countries."[22] But the "holding pattern" surrounding enlargement was evident here too: a survey I conducted in the summer of 1995 demonstrated that the debate had not begun in any European legislature nor had it begun in the Canadian legislature. In the Congress, which had, as noted, come forward with several bills on the question, the "Brown Amendments" to the 1994 NATO Participation Act had been considerably diluted by October 1995, despite their bipartisan sponsorship: references to the security of Central Europe being in the security interest of the United States and NATO, the naming of countries, and a mandatory reporting requirement upon the president to designate countries eligible for assistance had all been deleted.

The Polish-American Congress, in letters dated October 5, 1995, to members of Congress, blamed "an all-out effort" by the Department of State to "kill" the original legislation. It lambasted the NATO Enlargement Study as "purely theoretical with no practical signficance to the countries. . . . Contrary to the State Department's claims, countries which are not designated in the first stage of expansion will not be discouraged by the gradual but specific progress of NATO expansion. They will be discouraged—and are discouraged—by the total lack of any progress." Congress could not, of course, dictate enlargement, and the practical assistance program remained intact, but the political signals suggested that the mood in the Senate was perhaps geared less toward showing the way than it had been in 1994.

Other evidence of a leisurely approach to enlargement concerned the Clinton-Yeltsin meeting at Hyde Park on October 23, 1995. It occurred on the sidelines of the UN's fiftieth anniversary celebration, an occasion that Yeltsin had previously suggested could host an understanding on NATO enlargement. Former State Department official Peter Rodman alleged that Clinton had assured Yeltsin that if Russia cooperated with NATO in Bosnia, enlargement would "remain on the back burner," possibly even into a second Clinton term.[23] Although any such "arrangement" was denied by Secretary Christopher, it was interesting that no major U.S. statements on enlargement appeared for the rest of the year.

8

Intensified Dialogue

On December 5, 1995, the NATO foreign ministers in Brussels decided a three-point second phase program for 1996. The program, a continuation of the limited "how" and "why" approach, would

• pursue "intensified dialogue" through bilateral and multilateral consultations, with partners building on the foundation of the enlargement study;
• *strengthen the PFP*, which for some partners "will facilitate their ability to assume the responsibilities of membership"; and
• consider what *internal adaptations* are necessary to ensure that enlargement preserves the effectiveness of the alliance, particularly resource and staffing implications.

Through the intensified dialogue, interested partners would have the opportunity to learn more about the details of alliance membership, and the alliance would learn what partners could or could not contribute. "Participation in this next phase would not imply that interested Partners would automatically be invited to begin accession talks with NATO," and the phase would continue for the entire year with progress and the way forward assessed at the December 1996 Ministerial—where, as one NATO defense minister put it, "the decision to decide" would be made. The spring would be given over to political issues such as settlement of territorial or minority disputes, the summer to military questions, and the fall to a summing up.

A senior Polish official observed a few days after the NAC meeting that "1996 will be another dead year." Others preferred a "grace period" as a favor to Yeltsin on the (wholly unsubstantiated) assumption that "toning down the enlargement rhetoric" would somehow help the Russian president's election prospects (even though, as discussed earlier, it proved to be Russia that began floating compromise ideas months before the July 17, 1996, first round of voting). This Polish diplomat confided, however, that he had reasons to believe that Poland would become a NATO ally by the time of the fiftieth anniversary of NATO in April 1999, and new Polish president Aleksander Kwasniewski stated, before the NAC on January 17, 1996, that he expected phase two to "culminate with the date being set, before the end of this year, for the beginning of the accession talks."

On January 29, 1996, Ambassador von Moltke met with partners to explain that each interested partner would be invited to prepare "discussion papers" of much less detail than the PFP IPPs (Hungary was the first, on January 29, to request such talks, whereas the Czech Republic was the first, on March 14, to present an aide-mémoire). The papers would propose how to address the issues raised in the Enlargement Study (for example, policy toward foreign troops and nuclear weapons, relationship to the integrated command). Individual and informal talks would take place between NATO teams and partners addressing specific fields such as budgets, intelligence sharing, minimum interoperability standards, and general defense planning—not just relating to forces governing peacekeeping. A progress report would then be presented to the NAC in June in Berlin, with a final report submitted, later in December, to the NAC in Brussels.

By this time 11 countries had applied for NATO membership: Albania, the Czech Republic, Estonia, Hungary, Latvia, Lithuania, Former Yugoslav Republic of Macedonia (FYROM), Poland, Romania, Slovakia, and Slovenia. Except for FYROM, all of them were also participating in the PARP—as were Austria, Finland, Sweden, and Ukraine—to enhance interoperability with respect to PFP objectives. (See appendix D, this volume.) In 1996 NATO also made available its own defense planning questionnaire to partners to familiarize them with how allies decide what national forces are available; this went well beyond PARP interoperability objectives for limited PFP purposes to cover all forces and readiness.

Of course, the participation of 14 partner countries in IFOR, under NATO command (with the exception of Russia), itself would provide a wealth of experience. Indeed, it was even estimated that *three months in IFOR would be worth three years of PFP*.[1] On the PFP side, the momentum of exercises, including their broadening to include command post exercises, led the vice chairman of the U.S. Joint Chiefs of Staff, Admiral William Owens, to record on October 6, 1995, on the occasion of *Cooperative Engagement* in the Czech Republic, that "there has been a transition from simply counting the number of exercises to the substance of these exercises. And who knows the specific ways in which this will be used in future."[2] (In this context, recall Polish general Wozniak's observation in chapter 4, this volume, about the progression from peacekeeping to peace enforcement.) Indeed, IFOR partners were already undertaking "peace implementation" with NATO, and this explained why Poland chose to operate with U.S. commanders in Tuzla rather than under French command in Sarajevo. At 16 events, the 1997 PFP exercise schedule exceeded 1994's more than five times. It increasingly emphasized the important category of command post exercises to drill procedures among staffs and increasingly attracted officers instead of primarily conscripts, whereas in December 1995 "peace enforcement measures" were added to the roster of PFP topics and activities.

Would Russia eventually prove an interested NATO partner? Russia did not sign the NATO-related annex to the November 1995 Dayton Agreement, and its three light infantry battalions were under the operational control of a U.S., but not "NATO," divisional commander, despite public relations efforts. The official symbol of IFOR, for example, contained IFOR in Russian letters on the right side, although the actual translation is *sili volpanyeniya saglashyeniya*, or CBC. Foreign Minister Kozyrev candidly revealed the consequences of his government's insistence on pursuing Russia's *osobi kharaktyer* ("particular character") almost for its own sake with *de rigueur* opposition to NATO. Russia, he said, was losing time instead of developing special relations with NATO that could prove an alternative to NATO enlargement. He referred to a September 1995 NATO proposal, not made public, that fleshed out the "summary of conclusions"—to which Russia had not responded—of a "comprehensive program for consultations on such issues as non-proliferation, terrorism and definition of new challenges." His Ministry's proposals continued to "encounter categorical objections and [were] blocked in the process of coordination" among those favoring cooperation, non-cooperation, and confrontation, with the result that there was "no policy at all on NATO" within the Russian government.[3]

A narrowing of unpredictability seemed unlikely, particularly in the run-up to the June presidential elections that saw Yeltsin trailing behind Communist Party chief Gennadi Zyuganov, whose party had captured the largest bloc of seats in the new Duma following the December 17, 1995, elections. Seeking to capture votes from the Communists and "Liberal Democrats," in early 1996 Yeltsin purged his government of "reformers," including Kozyrev. Yeltsin made a point of holding a press conference detailing his telephone call to President Clinton on January 26, 1996, during which the Russian president stressed opposition to NATO enlargement, which could do harm "to Russia, Europe, and maybe the whole world."[4]

Concurrently, Russia continued to pursue the line, now more than four decades old, favoring a pan-European security system. On January 19, 1996, in the OSCE discussions in Vienna on a "security model" for the twenty-first century, a Russian proposal accepted at the December 1994 OSCE summit in Budapest, Russian head of delegation and ambassador Vladmir Shustov asked: although it was an OSCE principle that states had the right to belong to alliances, "is it expedient to realize this right" if it strengthens the security of some at the expense of others? Shustov urged that a document of "historic" proportion be adopted later that year at the Lisbon OSCE Ministerial that would outline a future European architecture; given the controversial nature of this notion—for example, concerns that Russia would seek control over NATO through some new directorate—Russia could always lobby for prolonging the model discussion and delay enlargement.

In anticipation, the NATO Enlargement Study stated that discussions on the security model should "reflect" but not "delay" the process of NATO enlargement, and the head of the Polish OSCE delegation Jerzy Nowak countered on January 26 that "nothing . . . can or may be construed as preventing *de facto* or indirectly extending existing treaties of Alliance." Nowak rejected the Russian slogan, "first the Model, then the rest," and dismissed the Russian notion that joint NATO-Russian "cross guarantees" to Central and Eastern Europe as an alternative to NATO membership would invite the creation of an unequal, marginalized, and subordinate security status for the Central and Eastern European nations tempting great power rivalry. For example, in September 1938 Soviet foreign minister Maxim Litvinov pledged that the USSR would assist Czechoslovakia against aggression provided France did likewise; the Munich capitulation followed the week thereafter without the USSR, let alone Czechoslovakia, having been consulted.

Warning Signs

The allies did not project a unified view. As UK defense secretary Michael Portillo admitted on January 23, 1996, "I don't see this as a great ambition of NATO to get bigger. What we are responding to is applications from other countries. We must take the time that is necessary to make wise decisions." He went on, "There can be no world security without taking Russia into account."[5] In my view, this testified yet again to a holding pattern approach and a kind of Russian veto. In contrast, the next month Secretary Perry suggested that more should be done to maintain momentum: "NATO enlargement is inevitable, and if NATO enlargement is the carrot encouraging reforms, then we cannot keep that carrot continually out of reach."[6] Nevertheless, Alexander Vershbow, senior adviser for Europe on the National Security Council, stated on January 31, 1996: "We will be coming up to the edge of a decision on the who and when, but I cannot predict at this point whether [sic]—exactly when that decision will be taken. . . . I simply wouldn't want to predict what the ministerial in December . . . will decide as far as what next."[7]

Reflecting a not-so-tacit Western agreement to "tone down" enlargement "rhetoric" on the theory that this would help Yeltsin in the elections (even though the desire to support Gorbachev had not prevented Bonn and Washington from insisting upon rapid German unification in NATO), Chancellor Kohl stated in Moscow on February 19, 1996, that enlargement "is not a topic which should be discussed precisely at this moment in time."[8] The chancellor went even further by citing the presidential elections in the United States as another reason why "an issue like NATO's expansion cannot top the agenda" and asserting that what was required, even *six years* after the fall of the Berlin Wall rapidly defined a new geopolitical status quo oriented toward the

West in the region, was "time to discuss the best ideas" to ensure that Russia's concerns were accommodated.[9] On February 13, U.S. deputy secretary of state Strobe Talbott insisted, "We must resist the temptation to accelerate the enlargement process for certain countries"; if the United States did so unilaterally, "it could well jeopardize our ability to maintain the necessary consensus among our NATO allies."[10]

Russian foreign minister Kozyrev, who was accused of an overly pro-Western and concessionary bent, had stepped down in early January 1996. Nevertheless, a few days later the new Russian foreign minister Yevgeni Primakov, who was associated with the past, stated that although he took a negative attitude toward NATO expansion, it was particularly important to avert any moves that could bring "NATO's military infrastructure closer to our territory."[11] On March 1, Primakov went further by saying that the time had come to find a solution. (The presidential elections were still in the distance and thus seemingly out-of-sync with the Western logic of avoiding discussion.) "I would have thought that one criterion for a possible compromise," he said, "would be not advancing the military structure of NATO to our borders."[12] This signal was somewhat confused by his also suggesting that the Central and Eastern European nations obtain unilateral NATO, or "cross" NATO-Russian, security guarantees without joining NATO, or opt for "political membership" in the alliance. Yet on March 21, a Russian Foreign Ministry spokesman referred to a "joint search for appropriate solutions," and in Oslo on March 25 Yeltsin himself advised states striving to join NATO "to do it like the French: becoming a member of the political committee without joining the military organization"—an option he had suggested in December 1994 could apply to Russia as well (and which Gorbachev had proposed in May 1990 for a united Germany).[13] Although on the same day Yeltsin's press secretary tried to argue that the president was referring to cooperation within the NACC and PFP, this is clearly not what Yeltsin had said.

In any event, as Robert Zoellick, former U.S. undersecretary of state and key player in the 2 + 4 negotiations during the Bush administration, had testified on April 27, 1995:

> I am skeptical that the fate of Russia's reform depends on whether NATO expands. . . . In addition, given the great un-

certainties about Russia's political future, it would be a mistake to try to fine tune our policies to suit the twists and turns of Russia's internal debates.

It certainly should not be surprising that Yeltsin, Kozyrev, and others have toughened their rhetoric about NATO expansion as the U.S. and others have signaled their uncertainty. It *is* surprising that we should shrink from pursuing our own interests, and those of our non-Russian friends, when some Russians have tried to transform their weakness into a threat [a staple of Russian diplomacy] . . . if we back down, the next time the hard-liners have a contest with moderate Russians, the hard-liners will be able to argue that sternness with the West pays off.[14]

At the spring NAC on June 3, 1996, in Berlin, the ministers pledged that "enlargement is on track." They also directed that the PFP be strengthened for the longer term by increasing partner involvement in shaping PFP programs, planning exercises, and other activities through the PCC and with the major NATO commanders and subordinate commands; by developing the IFOR experience through PFP regional cooperative programs; and by participating in CJTFs intended to facilitate the participation of "coalitions of the willing" in operations, including peace support operations outside the NATO treaty area. And yet hurdles remained:

• French foreign minister Hervé de Charette declared that France attached the "greatest importance" to a decision "on at least the principles underlying the reform of NATO and the general direction it will take *before* enlargement."[15] The politically difficult process of restructuring, which entailed cuts in NATO's 65 headquarters and reorientation toward missions such as IFOR, would doubtlessly prove difficult. This restructuring effort had been announced in 1990 but had made slow progress; indeed, the challenges—for example, the task of drawing France closer to an adapted alliance—could delay enlargement indefinitely. In Berlin the ministers declared that "the overall adaptation of the Alliance will facilitate" the process of enlargement, but this adaptation could also further stretch enlargement decisions out into the future.

• Reflecting concerns about how to reassure partners not invited in the first round, NATO assistant secretary general von

Moltke stated, "Our aim is to narrow the distinction between Allies and Partners as much as practically possible."[16] Yet, might this lead to just another installment of NATO non-differentiation, considering that "there will always be an unbridgeable difference" between collective defense and PFP?[17] Strengthening the PFP was inherently desirable, but then NATO was far more than the PFP. At the December 6, 1995, NACC ministerial, Poland again felt compelled to request a "variable geometry" strategy implying, it argued, tailored pre-accession programs so as not to waste resources. Indeed, the greater the knowledge of NATO procedures the more candidate nations would also be able to participate in NATO's new missions, as Polish diplomat Robert Pszczel argued:

> For a non-NATO country such as Poland the difficulty of ensuring human and technical compatibility with NATO forces severely limits the capacity to field a larger number of units [than Poland's one battalion in IFOR] . . . it is membership alone that will provide the full interoperability standards for at least the bulk of a nation's armed forces.[18]

By 1996, for instance, 800 of the more than 1,500 NATO STANAGs and other allied publications were still not releasable to partners, and at the June 4, 1996, NACC meeting, Polish foreign minister Dariusz Rosati again warned that the *priorities* of allies and partners had to be "clearly spelled out" to make cost-effective decisions. In mid-July Lithuania called for NATO to announce at its December 1996 ministerial the creation of an "Atlantic Partnership" for all the states (eleven) striving for eventual NATO and EU membership—*not each and every partner*—who would enjoy intensified PFP activity aimed at enlargement, such as participation in NATO collective defense (Article 5) exercises on Baltic state territory. This was subsequently endorsed by all three Baltic states as a "Partnership for Security" by which NATO would formally identify "applicant partners."

• Ukraine, the only other country besides Russia that enjoys a special relationship with NATO outside the PFP, and which NATO had declared was playing a "growing role in European security," also factored in the reassurance policy. Unlike Russia, Ukraine did not oppose NATO enlargement, but expressed concerns that it would become a "buffer zone" vulnerable to Russian pressure as a result of enlargement. Thus, it sought a guarantee that new members would have nuclear weapons-free status. Ukraine had suggested at the Berlin NACC on June 4, 1996, that its nonaligned status might change and by the fall was even suggesting that NATO agree to legally binding guarantees of Ukrainian security. (In December 1994, Russia, the United Kingdom, and the United States had concluded a "Memorandum on Security Assurances in Connection with Ukraine's Accession to the Treaty on the Non-Proliferation of Nuclear Weapons" with Ukraine, in which they essentially restated the UN principle of refraining from the threat or use of force and to seek "immediate action" in the UN Security Committee should such a threat or act of aggression occur involving nuclear weapons.) Polish deputy defense minister Andrzej Karkoszka argued, however, that "we can only accept full membership" and "cannot say in advance" that "some specific weapons on our soil" should be excluded.[19] This meant that new members might not wish to host nuclear forces, or any stationed forces, in peacetime, but that such arrangements should not be preconditions for NATO membership, just as they had never been for current allies.

• In a September 6, 1996, address in Stuttgart, billed as the Clinton administration's "vision" for a "New Atlantic Community" for the twenty-first century, Secretary Christopher broke important new ground: "We should invite several partners to begin accession negotiations" at a summit in spring or early summer 1997, while also concluding a formal NATO-Russia "Charter" (a term first raised by French president Jacques Chirac on June 8, 1996) for standing consultative arrangements and joint action in conflict prevention, counterproliferation, and environmental disaster prevention.[20] On September 2, however, Secretary of State Rifkind referred simply to "the possibility of beginning talks" about enlargement,[21] and Germany and France reportedly sought to conclude the NATO-Russia Charter *before* enlargement decisions were made, thus putting the ball back into Moscow's court.[22]

• During his October 7, 1996, visit to NATO, then Russian Security Council secretary Aleksandr Lebed, positioning himself to succeed President Yeltsin albeit dismissed from his post on October 17, pledged closer Russian cooperation in the PFP (the carrot). He called for a *sequential* process for NATO's finalizing of its reform plans, a formal NATO-Russian treaty to regulate joint decisionmaking, and only then a decision on enlargement— a decision that, he urged, should be delayed until "the next generation."[23]

• A further potential hurdle involved public opinion in the candidate nations themselves. Even if the political elite were to favor NATO membership as soon as possible, U.S. Information Agency polls sponsored in 1995 showed that Polish opinion alone agreed, by a slim 55 percent percent majority, to implementing the basic alliance contract of coming to the defense of other countries. Yet, opinion in Poland as well as in Bulgaria, the Czech Republic, Hungary and Slovakia—with Hungary, the Czech Republic, and Poland as the top three targets of foreign direct investment—opposed by large majorities the increasing of national expenditures toward military (versus social) needs.[24] All of the former Warsaw Pact nations, however, had experienced steep declines in military budgets and combat effectiveness, thus raising the question posed by Joseph Kruzel as to what exactly they would be able to *contribute* to alliance security.

Foreign Minister Primakov, nevertheless, again suggested in July 1996 that "a way out of the prevailing situation might be found through compromise and reciprocal consideration of interests."[25] They must include a "guarantee that the enlargement of NATO will not lead to the installation of military infrastructures"[26] and that a NATO-Russia treaty will not be reduced to a mere declaration, as he suggested some in the alliance sought, but should contain guarantees to allay Russian security concerns.[27] Even so, however, Russia sought to "cap" enlargement after the first new members were invited, a view not without resonance within the alliance.

To what extent this was the "Russian government" position was unclear, for at the end of 1996 Russian defense minister Igor Rodionov dismissed NATO's new role in peacekeeping as a "sham" covering "the bloc's continuously growing military potential." Instead of enlargement of an alliance, NATO should transform into a peacekeeping organization under UN and OSCE auspices.[28] Additional interesting Russian perspectives were provided by new Security Council secretary Ivan Rybkin, who on October 30 proposed that Russia become a political member of NATO.

Rodionov's ally Lebed, who had once predicted that NATO enlargement would mean "World War III," now stated that if NATO had "enough money and energy to expand, feel free."[29] Time would judge the energy requirement, but in March the U.S. Congressional Budget Office (CBO) released a study indicating that the costs of enlargement were hardly unmanageable. Although dated because it included Slovakia with the Czech Republic, Hungary, and Poland, the study estimated that from 1996 to 2010 costs would range from $60 billion to $125 billion. The lower figure entailed the costs of enhancing indigenous forces and preparing reinforcement reception facilities, of which new allies would cover two-thirds of the cost; the higher figure took into account stationed forces.[30] Because the larger amount would presumably be demanded only with the rise of a neo-imperial Russia, and the CBO study was indeed based on this worst-case assumption (sending no doubt an unhelpful political signal at a time when NATO was trying to persuade Russia that it was not the target of NATO enlargement), it would seem that Russia would have a vested interest in not compelling NATO to invest resources to hedge against a new Russian threat. A NATO posture aimed at force projection to accommodate new members would also perhaps prod allies to move away from static Cold War-era defense postures and thus make an important contribution to "burdensharing."[31]

On September 4, 1996, Primakov met in Bonn to discuss a German initiative for launching the elusive comprehensive NATO-Russian dialogue, whereas at the NAC meeting in Berlin it had been agreed that Russia could assign permanent liaison officers to NATO and NATO post officers in the Russian General Staff (an offer on which Russia had not moved). Richard Holbrooke suggested that NATO-Russia cooperation did "not mean that NATO should include Russia, as once was suggested," but instead proceed in "perhaps some new structures which can be a result of 16 + 1 dialogue."[32] German foreign minister Kinkel favored changing this formula to "17," as Russia preferred, to connote a status other than invitational in alliance deliberations.

In any event, on September 26, 1996, Secretary Perry offered Russia a five-point program of cooperation at the informal NATO defense ministers' meeting in Bergen, Norway:

• Consultations for a possible follow-on force to IFOR, rather than once again having to accept an operational plan already settled by the allies;

• establishment of permanent military liaison officers (already offered during the Berlin NAC), which Lebed had stated in October he would support;

• Russian participation in CJTF planning;

• a "mechanism" for crisis consultation (already part of the PFP); and

• the scheduling of regular meetings with Russian officials along with corresponding NATO officials.

The NATO Enlargement Facilitation Act

This bill stands for the proposition that neither we nor the emerging democracies of Central and Eastern Europe can afford to wait any longer. Only by taking this step now can we ensure that the democratic gains of the last seven years are not going to be reversed. After today's vote, it is hoped that we will never again hear that the Congress does not support NATO enlargement.[33]

—The Honorable Benjamin Gilman
Chairman of the International Relations Committee
U.S. House of Representatives
July 23, 1996

The image projected during the year of "intensified dialogue" was still that the West simply lacked a definition of success for its enlargement policy. "Today we still cannot say," Polish deputy defense minister Karkoszka stated in June 1996, "that we are 100-percent certain that our membership in NATO is a foregone conclusion."[34] Thus, Congress again acted.

The NATO Enlargement Facilitation Act was passed on July 23, 1996, in the House by an overwhelming majority of 353 to 65 and in the Senate on July 15 by 81 to 16. Its key provisions were as follows:

- "The United States continues to regard the political independence and territorial integrity of all emerging democracies in Central and Eastern Europe as vital to European peace and security";
- The Congress finds that Poland, Hungary, and the Czech Republic have made the most progress toward achieving NATO eligibility criteria (the Senate added Slovenia, accepted in September in conference between the two chambers);
- An amount of $60 million was authorized to facilitate the transition to NATO membership, with the four countries designated as eligible to receive assistance;
- The process of enlarging NATO should not stop with the admission of these countries, and the president may designate other countries as eligible, including Albania, Bulgaria, Estonia, Latvia, Lithuania, Moldova, Romania, Slovakia, and Ukraine, whereas the countries of the Caucasus region should not be precluded from future NATO membership.

The act did not set any dates for membership and did not identify any country clearly as an early candidate in so many words. Nevertheless, it again sent a *clear legislative signal* from one of the NATO national legislatures that would have to approve ratification in the Senate and funding in both chambers.

The White House welcomed the legislation as sharing the administration's "determination" to enlarge NATO, and President Clinton signed the act on September 30, 1996. Both the Democratic Party and Republican Party election platforms supported NATO enlargement—the Democrats "in the near future" and the Republicans "immediately." On October 22, 1996, President Clinton announced "America's goal" as follows: "By 1999, NATO's 50th anniversary and ten years after the fall of the Berlin Wall, the first group of countries we invite to join should be full-fledged members of NATO."[35] On November 21, the NAA welcomed the passage of the NATO Enlargement Facilitation Act and endorsed the proposal drafted by Senator Roth for the alliance to conduct annual bilateral consultations to review *all* candidacies for membership.

Yet, looking ahead, UK ambassador John Goulden called that same month for a greater sense of *priorities* to manage alliance business:

More money and personnel need to be shaken out if we are to cover the new priorities. . . . We have too little money for partnership, and none allocated yet for the costs of enlargement. . . . Is it sensible to envisage a partnership embracing all who wish to take part, in which Partners have all the benefits of full membership except the defense guarantee and the right of decision-making in the Council? If so what safeguards would be needed to ensure that NATO preserved its core assets as a military alliance. . . . Is there a limit to membership, beyond which we would cease to be a serious defense Alliance?[36]

9

Endgame Afoot

On December 10, 1996, the NAC convened in Brussels. The time to discuss the way forward had now arrived, and a more specific outreach strategy for NATO enlargement, enhanced PFP, and special NATO relationships with Russia and Ukraine was adopted.

One or More

First, the allies accepted President Clinton's proposal for a summit in 1997, setting a firm date of July 8–9 in Madrid (Spain having become the most recent NATO ally), at which enlargement would, the ministers pledged, at long last be launched:

> We are now in a position to recommend to our Heads of State and Government to invite at next year's Summit meeting one or more countries . . . to start accession negotiations with the Alliance. Our goal is to welcome the new member(s) by the time of NATO's 50th anniversary in 1999. We pledge that the Alliance will remain open to the accession of further members in accordance with Article Ten of the Washington Treaty.[1]

No countries were mentioned, but in the U.S. Senate on February 5, 1997, the new NAA president, Senator William V.

Roth, Jr. (R-Del.), submitted a Concurrent Resolution calling upon the alliance to extend invitations to nations "including" the *Czech Republic, Hungary, Poland,* and *Slovenia*—the four countries that were already identified in U.S. legislation. Seeking to incorporate a major Southern European nation in the first wave, France, Greece, Italy, Spain, and Turkey sponsored *Romania* (the second largest nation in this group of five in terms of population, territory, and armed forces). As then U.S. secretary of state Christopher had stated on July 17, 1996, Romania (which would see the coming to power of a reform government under President Emil Constantinescu the following November, replacing the post-Ceaucescu ex-communist government) had accomplished "a great deal to qualify itself for early consideration for membership."[2] As long ago as 1967, Romania had essentially adopted only political membership in the Warsaw Pact. It had an early start on diversifying its sources of such critical items as air defense and was the first Central European nation to either accept such major U.S. systems as the C-130 *Hercules* transport aircraft or co-develop them (the *Cobra* armored helicopter). I also lent my support to Romania in the first wave, as I informed Prime Minister Victor Ciorbea in Bucharest on May 27, 1997. By that time Romania had concluded treaties with its neighbors Hungary and Ukraine to resolve territorial or ethnic issues, and the congressional Republican leadership took this position as well.

Given mounting transatlantic concerns about the course of democracy under Prime Minister Vladimir Meciar, *Slovakia* was regrettably no longer regarded as an early candidate. The fact that Slovakia geographically connected Hungary and Poland to the Czech Republic was simply insufficient to overlook these problems. (Moreover, Slovenia, which had served as the planned axis of the main Warsaw Pact attack on the southern NATO region, would link Hungary with Italy.) The 1996 U.S. Department of State Human Rights Report found "disturbing trends" away from democratic principles, and on April 29 Meciar was informed by congressional leaders of the Commission on Security and Cooperation in Europe (the Helsinki Commission), Senator Alphonse D'Amato, Republican chairman from New York, and Representative Christopher H. Smith, Republican co-chairman from New Jersey, that Slovakia's human rights record "most probably" took it out, in their view, of the first wave of new alliance members. Domestic political turmoil even caused a referendum, held May 23–24 on the issue of NATO membership, to fail because of insufficient participation. It involved a boycott by supporters of pro-Western president Michal Kovac and led to the resignation of Foreign Minister Pavol Hamzik on the grounds that the failure to send a political signal greatly limited his ability to promote the foreign political priorities of Slovakia. On May 28 Italian defense minister Beniamino Andreatta affirmed, however, that geostrategic interests favored Slovakia's membership in NATO, but not, he stated, until a second round.

Thus, the range of "three-to-five" was commonly mentioned by NATO officials early in 1997. Germany, the United Kingdom, and the United States reportedly gravitated toward "three"—the Czech Republic, Hungary, and Poland. The new UK foreign secretary under the Labour government Robin Cook acknowledged on May 19 the "very strong" claims of those three nations, but also stated there was still no "final view" on Romania and Slovenia. By this time the admission list stood at 12 countries, with the addition of Bulgaria (which had elected a reform parliament in April 1997).

As explicit reassurance to Russia, the NATO foreign ministers also adopted the so-called three noes: "NATO countries have no intention, no plan, and no reason to deploy nuclear weapons on the territory of new members nor any need to change any aspect of NATO's nuclear posture or nuclear policy—and we do not foresee any future need to do so." Another unilateral NATO statement of March 14, 1997, declared that in the foreseeable security environment NATO would carry out its collective defense and other missions by ensuring interoperability, integration and capability for reinforcement rather than by "additional permanent stationing of substantial combat forces." Both statements built on the relevant language of the 1995 *Study on NATO Enlargement*.

In arms control negotiations, these pledges were enhanced. On February 20, 1997, in CFE, NATO proposed eliminating the anachronistic "group of states" structure of the 1990 treaty (NATO and the Warsaw Pact) in favor of national and territorial ceilings (the latter covering both national and stationed equipment in the CFE zone, which does not include the territory of Canada, the United States, and regions of Russia and Turkey), and lower force levels including steps by alliance nations to reduce ground equipment "significantly" below the *current* group ceiling. NATO also proposed a stabilization measure whereby territorial ceilings for total ground equipment would be set no higher than current national levels in the area of Belarus, the Czech Republic, Hungary, Poland, Slovakia, parts of Ukraine, and the Russian region (*oblast*) Kaliningrad, so that when NATO enlarged it would not pose a new concentration of higher force levels. Exceptions for exercises, peacekeeping, and temporary deployments would be allowed. In the OSCE, NATO proposed a transparency measure on April 16 that would require annual reports on new military infrastructure, or substantial improvements to existing infrastructure, subject to verification of military airfields, storage facilities, fixed air defense sites, headquarters, et cetera.

These undertakings and proposals fell short of Russian calls for a complete ban on the permanent stationing of all ground and air CFE-limited equipment outside the territory of present NATO members, collective limits on alliances, and binding "guarantees" against infrastructure upgrading and the deployment of nuclear weapons. By late April, however, Russia had dropped its demand for alliance ceilings and was exploring such variations as limits on the percentage of foreign forces and infrastructure that could be hosted in a state party, albeit at very low levels. (For example, the limit—no more than 1.5 percent of tanks or 5 percent of manpower—kept Poland from hosting more than a brigade's worth of other NATO forces.)

Nevertheless, NATO was endeavoring to reassure Russia that enlargement would not pose an objective military threat or concern. At the same time, the alliance was signaling that new members would not be forced to accept "second class" security status in NATO. As Adam Kobieracki, deputy head of the Polish OSCE delegation, insisted on March 11 that "under no circumstances" could CFE Treaty adaptation preclude the right of any state to join NATO and to station NATO troops *permanently or provisionally*. NATO assistant secretary general von Moltke explained the alliance approach on April 28, as follows:

> Russian perceptions do matter and they must be taken seriously. Continuing Russian anxieties are based on a profound misunderstanding of NATO's character and intentions. All the more reason, therefore, to make a special effort to allay those fears and remove the misunderstandings. But this cannot be done—and will not be done—at the expense of other European countries and their interests.[3]

In what might be seen as yet a third unilateral statement, Ambassador von Moltke also added: "There will not be the large-scale build-up of new infrastructure as this is not needed and resources will be limited. Many of the infrastructure left by the Warsaw Pact will not be used. . . . " Von Moltke noted that of 27 military airfields in the former GDR, only three were operational today, the rest having been abandoned or privatized. In Poland only 16 percent of this infrastructure—of nearly 2000 former Soviet installations—was being used.

At the same time, NATO would require the use and upgrading of new member infrastructure. The alliance had pledged reliance on interoperability, integration, and reinforcement rather than on substantial stationed combat forces as well. For example, the United States had spent $10 million to upgrade the Taszar airbase in Hungary to serve as a staging area for IFOR/SFOR (covering diverse areas such as telephone upgrading, runway repair, water chlorinators, gravel and asphalt work) out of around $20 million unilaterally invested in Hungary to support the operation.

The NAC also requested analysis of the "relevant factors" associated with admission, recommendations on adaptation of alliance structures, and preparations for a plan for conducting the accession negotiations. On December 17, the defense ministers also requested study of the military factors associated with accession and the financial implications. On May 29, 1997, little more than five weeks before the summit, the NAC would convene again in Sintra, Portugal, when last-minute decisions about whom to invite, coupled with other difficult decisions about alliance restructuring, might have to be made at the highest level.

Enhanced Partnership

NATO was also keenly interested in ensuring that partners who would not be invited early, or those who did not wish to join, would remain interested in the PFP and in rationalizing the NACC and PFP to combine the best of both worlds. The NAC in Brussels returned to similar themes sounded at its last meeting in Berlin by offering these areas for enhanced cooperation, even if some of what was offered was already being undertaken in the real-world IFOR/SFOR operation. Cooperation would

- enhance the political dimension of the PFP;
- expand PFP missions to the full range of the alliance's new missions (excluding collective defense) "as appropriate";
- broaden the PFP exercise program accordingly;
- enable partners to participate in the planning and execution of PFP exercises and operations;
- involve partners more substantively and actively in PFP-related parts of the regular peacetime work of NATO military authorities;
- enable partners who joined NATO-led PFP operations to contribute to political guidance over such operations;
- examine possible modalities for a "political-military framework for PFP operations," building on the PMSC;
- enhance partner participation in PFP program decision-making;
- increase regional cooperation;
- expand PARP to move beyond the PFP *interoperability* objectives toward *force* planning akin to that practiced by full NATO members, such that eventually the only difference between a full member and partner would be the Article 5 pledge;
- establish partner diplomatic missions (not just liaison offices) to NATO, with the relevant Belgian law having entered into force in March 1997; and
- work with partners on the initiative, proposed by then U.S. secretary of state Christopher in September 1996, for an "Atlantic Partnership Council" to provide for "greater coherence in our cooperation."

By April the name had become, owing to Russian initiative, the Euro-Atlantic Partnership Council or EAPC. What improvements it would bring had to be determined, for the list of topics as settled by May was identical to those of the NACC. It was unclear *how* partners would play a greater role in decision-making, but ostensibly the alliance had recognized that the NACC could not endure forever, particularly as nearly all partners were far more interested in the nuts and bolts of the PFP and in more focused political discussions than the NACC offered. As a NATO report of December 3, 1996, *Review of NACC at Five Years*, concluded: "The situation in the NACC area is now radically different from that of the time that NACC was created."[4] In general, the NATO Senior Level Group on PFP Enhancement concluded by May that the EAPC would aim to (1) strengthen political consultation; (2) develop a more operational PFP role; and (3) provide for greater partner involvement in PFP decision-making and planning. As the successor to NACC, the EAPC would provide the overarching framework for consultations on a broad range of political and security-related issues, with the enhanced PFP an element within the framework, and, in short, would develop and consolidate NATO outreach.

On April 3, the Military Committee completed its "Concept for PFP Enhancement," its military advice to NATO political authorities. It recommended Permanent Staff Elements at NATO headquarters open to all partners (whether or not seeking membership), the granting of partner authority over NATO/PFP exercises on their territory, partner positions in NATO international staffs, additional personnel and funding resources as the exercise program was already stretched to the maximum, more availability of NATO documents, greater focus on the quality rather than quantity of exercises, and improved capability for direct and rapid information exchange within the PFP community. Interestingly, the foreign ministers had been generous with new directions, but the Military Committee was compelled to recommend "emphasis on austerity to limit the growth in demand" given resource constraints.[5]

These measures did not escape controversy. For example, although the intent of the enhanced PFP was to involve partners more in decision-making, General Klaus Naumann, the chairman of the NATO Military Committee, informed the NAA on February 16, 1997, that "we need to protect NATO's right to decide at 16. To embark on solutions which would necessitate consultation with Partners before NATO decides whether it wishes to act on its own would give all of us some difficulties to conduct efficient crisis management."[6]

France questioned the relationship between an EAPC of possibly 44 states and the OSCE of 54 states. The Netherlands favored a close link so that the EAPC could be seen as the security arm of the OSCE—the elusive "operational role" that had always eluded the NACC.

Because Article 5 obligations would come into effect upon

admission, sensitive NATO defense plans would have to be made available during the *pre*-accession phase. Yet, what if an invited nation did not survive ratification, either in the NATO national legislatures or in its own parliament? Would any of the present allies reduce the quantity or quality of intelligence shared with NATO as a result of national concerns about enlargement to former Warsaw Pact nations and individuals trained by Soviet intelligence services, even during a "nonadversarial" era?

Would common offices for partners, invited states, and full members be the exception or the general rule?

How could the NATO staff, under an increasing burden, simultaneously pursue accession negotiations, accommodate those not invited at an early stage, engage those nations not seeking membership, facilitate the special relationships with Russia and Ukraine, and pursue the "Mediterranean dialogue" with six North African and Middle Eastern nations? Already there were 250 political and military topics in the PFP work program; how would it be enhanced qualitatively?

"Neutral" countries such as Sweden did not wish to be perceived as participating in a new forum viewed as only a waiting room for NATO membership. Nor did they wish to be assigned, under the "regional PFP" rubric, some indirect responsibility for the security of countries seeking to join the alliance.

On the other hand, there was also concern that the enhanced PFP/EAPC might become a "consolation prize" or "decompression chamber" for partners not invited in the first wave. Certainly, the three Baltic States were expressing increasing anxiety. On June 20, 1996, President Yeltsin wrote to President Clinton that Russia was willing to provide security guarantees to the Baltic States, or to do so jointly with NATO "if needed" (just as he had proposed cross guarantees enshrined in a NATO-Russia agreement in September 1993). "It is out of the question even a hypothetical possibility of extending NATO's sphere of operation onto the Baltic states. This perspective is absolutely unacceptable for Russia, and we would consider any steps in this direction as a direct challenge to our national security interests" and "destructive of fundamental structures of European security."[7]

Five days later, President Clinton declared, at a meeting in Washington with the Baltic presidents, that the "the first new members should not be the last."[8] The U.S. ambassador to Sweden declared on November 26, 1996, that for the United States the question was not "if" but "when" for "Baltic membership in NATO."[9] During the November 1996 NAA Annual Session in Paris, Norwegian MP Jan Petersen eloquently argued that his NATO nation had always shared a border with Russia, just as Poland and Lithuania did with respect of Kaliningrad, and that ambiguity on the part of the West, citing discussions over whether neutral Finland and Sweden could provide a security blanket for the Baltics or that EU membership could "compensate" the Baltics for delayed entry into NATO, could encourage adventurism in an unpredictable Russia.[10] In early May, German defense minister Rühe stated that the Baltic states would "definitely" become NATO members "at some point."

On February 6, 1997, the Danish Parliament, the Folketinget, passed a resolution similar to one it had adopted, 104 to 14, in November 1994, urging the government to ensure that the Baltic states would be accorded the *same membership possibilities* open to other Central and East European states.

In the same spirit, I introduced a Concurrent Resolution on January 15, 1997, recommending the integration into NATO "at the earliest possible date" of Estonia, Latvia, and Lithuania, noting the consistent progress made toward civilian control of their militaries (all three countries had to start from scratch since becoming independent in 1991; their parliaments had actually preceded the armed forces, and none of them deployed a single tank or combat aircraft), rule of law, the free market, and human and civil rights. (This support was expanded to include Romania in the European Security Act of 1997, which I cosponsored in the House on April 24, 1997.) The Baltic states were integrating their forces in a joint peacekeeping battalion (BALTBAT), trained under NATO standards and equipped by seven NATO countries plus Finland and Sweden, and had acquired operational knowledge by participating in the Nordic IFOR brigade. Why then, with such active foreign military support, was there obvious hesitation about Baltic membership in NATO? All three nations shared a border with Russia, but so too did Poland. The Baltic armed forces were only being consolidated, such that Secretary Perry had stated in September 1996 that these nations were "not yet ready" to undertake Article 5 responsibilities (although Iceland has no armed forces, and Luxembourg's forces are one-sixth the size of Lithuania's). Beyond this military criterion, however, was there some unwritten geopolitical factor—some red line—in determining qualified future NATO members?

The Baltic states adopted the position in early 1997 that "at least one" of them should be admitted to NATO in the first wave. On December 9, 1996, Latvian deputy secretary of state for foreign affairs Maris Riekstins informed an NAA delegation that "the 'open door' slogan in connection with NATO is not sufficient for the Baltic States."[11] The three nations also urged the alliance to launch an enlargement *process* in Madrid by recognizing prospective NATO members at the summit and providing them with a timetable for full membership. This approach *could* have diminished perceptions of "dividing lines" while still allowing for a small group of countries to be admitted by 1999. Yet, at the same time there was the agreed alliance view that specific invitations in Madrid would demonstrate that the enlargement process was beginning, without locking the door to anyone. Moreover, NATO naturally could not admit all states seeking membership all at once or raise expectations without careful regard for meeting membership criteria. Yet, at the end of March 1997, during a congressional visit to the Baltic states (nations never recognized by the United States as Soviet republics), I stated that "the only way NATO can remain as a meaningful defense treaty is by taking in all the nations which qualify" and that ideally all three Baltic states should join simultaneously.[12] I also stated that *Russia, as with any other nation emerging from former Soviet domination, could also join NATO* when and if it met alliance criteria. Nevertheless, what if, in a so-called one-shot enlargement urged by Russia but also somewhat supported in the West, NATO's doors were closed after the first candidates were admitted? (Former U.S. senator Sam Nunn had urged a pause of up to *10 years* after the first admissions in 1999, based on the assumption—with which I disagree—that Russia would not be accepted as a NATO member.[13]) In that case, the United States would have to reconsider its own NATO membership.[14] A democratic and purely defensive alliance cannot be based on double standards or anachronistic notions of spheres of influence.

The NATO-Russia Act

The NATO foreign ministers also moved beyond the offer of a "political framework" with Russia to propose reaching, by the time of the summit, "a document or . . . Charter" to "deepen and widen the scope of our current relationship and provide a framework for its future development." Drawing on the 1995 "areas of pursuance," NATO offered

- shared principles that would form the basis of the relationship;
- areas of practical cooperation in the political, military, economic, environmental, scientific, peacekeeping, armaments, nonproliferation, arms control and civil emergency planning fields;
- mechanisms for regular and *ad hoc* consultations; and
- mechanisms for military liaison and cooperation.

In a historically novel step for the alliance, NATO secretary general Solana was delegated to consult with Russia on behalf of the 16 allies on the "possibility" of such an agreement. On the following day, Russia agreed to pursue these consultations. In September 1996, however, Foreign Minister Primakov, to avoid being viewed as "negotiating with NATO," expressed reluctance to acknowledge Solana as his counterpart, preferring instead to deal with NATO nation ministers. Even while maintaining its "negative position" toward NATO enlargement, Russia sought at least "damage limitation"—a new approach worked out in late 1996—and recognition of its desired "equal partner" status. (On February 23 President Yeltsin described Russia's primary task as "delaying" and "eroding" NATO enlargement.)

Within a few weeks, the potential cooperation menu had grown to include the term "co-decision." During an NAC meeting in Brussels on February 18, 1997, new U.S. secretary of state Madeleine Albright proposed reaching "concerted decisions where possible and [seizing] the opportunities to launch joint actions"—for example, through a NATO-Russia "Council" and a joint brigade.[15] The secretary of state also urged that accession negotiations be completed by the end of 1997, and it was assumed that the U.S. Senate, to which other allies would naturally look, would begin the ratification debate by March or April 1998.

In March President Yeltsin summarized Russia's position as follows:

- Russia remains negative to plans for NATO enlargement and "especially to the possible eastward advance of the alliance," and a decision to advance "could lead towards a slide into a new confrontation."

• The NATO-Russia relationship was not "payment" for enlargement but a separate issue (although the Russian approach was obviously tackling both strands).

• The NATO-Russia document must be a legally binding treaty containing "clearly worded guarantees" of Russia's security regarding nonexpansion of NATO military infrastructure eastward and the nondeployment of foreign forces outside the territories where they are presently deployed. In addition, NATO's unilateral statement on nondeployment of nuclear weapons should be written into the NATO-Russia document as a permanent pledge.

• Reassurance should also be provided by CFE Treaty adaptation, including limits on alliances, which should be entered into the NATO-Russia document.

• Joint discussions should be initiated on "issues concerning NATO's transformation, especially since this process is developing slowly, and the aims of NATO's adaptation to the new conditions, declared in Berlin last year, are far from being implemented." Such discussions should also be initiated for "developing coordinated approaches to *all* issues of European security as well as for taking *decisions on issues involving Russian interests on a basis of consensus.*"[16] (In Berlin NATO actually *reaffirmed* its core function of collective defense, and although progress was made on CJTFs for *both* collective defense and collective security missions, NATO had never adopted Russia's proposal that the alliance become a purely political and peacekeeping organization.)

Additional issues not explicitly or publicly raised at the presidential level were a ban on NATO membership for the former Soviet republics, including the Baltic states, and the idea of a single limited enlargement—the latter contributing to the anxiety of certain partners who feared that those invited in Madrid would be the last admitted for a time well into the future, as Senator Nunn had proposed. Yeltsin and his spokesman heated up the atmosphere by questioning NATO PFP activities in the former Soviet Union, with Yeltsin having termed a PFP exercise (*Sea Breeze*), scheduled for summer 1997 in the Black Sea and the Crimea, as "against Russia's wishes."[17] Russia had been invited but had refused to participate. Such a view of the PFP clearly raised the issue of how the NATO-Russia document would make a difference.

At the Russia-U.S. summit held on March 20–21 in Helsinki, both presidents Yeltsin and Clinton agreed to disagree on NATO enlargement. Nevertheless, it was also decided "to minimize the potential consequences of this disagreement." A NATO-Russia document would

- be signed by heads of state and constitute an "enduring commitment," as NATO did not favor a formal treaty; Yeltsin stated that the document would nevertheless be sent to the Duma for ratification;
- provide for "consultation, coordination, and to the maximum extent possible where appropriate, joint decisionmaking and action on security issues of common concern";
- reflect and contribute to "the profound transformation of NATO, including its political and peacekeeping dimension"; and
- refer to NATO's unilateral statement regarding nuclear weapons.[18]

Conspicuously absent was a similar reference to NATO's recently announced policy on the nonstationing of combat forces. This issue, coupled with a persistent Russian desire to obtain concrete guarantees rather than record NATO unilateral statements, plagued the follow-on efforts primarily through the Solana-Primakov channel. Reflecting the "protocol" versus "summary of conclusions" discussions in 1994, form again played a role. Russia, having initially sought a treaty, called for an "act" (NATO preferred "charter") comparable to the 1975 Helsinki Final Act of the OSCE. NATO sought to have the NATO–Russia Council chaired only by a NATO official, albeit with a Russian deputy, rather than jointly chaired, as Russia wished. The extent to which the NATO-Russia document could reflect general arms control issues was also being discussed, even though the details would have to be solved in Vienna with the participation of all concerned parties, not just Russia and the sixteen allies. Interestingly, the plan for membership accession negotiations requested by the NAC in December 1996 was still not ready, although NATO insisted there would be no delay in enlargement.

Communist Party leader Gennadi Zyuganov called the Helsinki arrangements a "Versailles" for Russia.[19] On January 24 an "anti-NATO association" was formed in the Russian Duma comprising in a few weeks a majority of the deputies, who on April 24 described NATO enlargement as *Drang nach Osten*, or "March to the East." This raised the obvious concern that *whatever* was agreed, despite two and a half years of intermittent efforts, could be exploited by nationalist forces as a "capitulation," thus complicating political life in the same way the Russian government argued that unconditional NATO enlargement could. There was an absence of compelling evidence, however, that public opinion in Russia shared this view. The highly popular Russian politician Aleksandr Lebed predicted that the result would be an "empty document, meant for use inside of Russia" to save face for the Yeltsin government.[20]

In the West, too, the Helsinki meeting prompted queries of a different sort. Had Washington gone too far for no obvious reason? Henry Kissinger labeled the encounter a "fiasco," in part reacting to President Clinton's incredible remarks at a press conference describing NATO as having been a "mirror image" of the Warsaw Pact (precisely the opposite of what NATO sought to convey to Russia). Kissinger asked whether Russian acquiescence was being purchased at an "exorbitant price." The envisaged NATO-Russia Council would dilute the NAC's ability to consult confidentially and rapidly, conduct crisis management, and preserve a credible, integrated military structure. Kissinger argued that Russia should have every opportunity to participate in the construction of a new international system, but such participation did not require "*de facto* membership of NATO" and could have been pursued through the OSCE or in a permanent "contact group" of France, Germany, Russia, the United Kingdom, and the United States. He proposed that the council not be activated *until* enlargement had taken place, or that, at a minimum, applicant countries participate in its deliberations, as well as those of the NAC, prior to their admission as full NATO members.[21]

Former U.S. national security adviser Brent Scowcroft disagreed with the whole idea of a NATO-Russia document, arguing that as a defensive alliance NATO should not be portrayed as the counterpart of any country. Former General Scowcroft also favored what could become "in effect" a "permanent" pause to enlargement, arguing that unlimited enlargement "would destroy NATO as we know it"—even though, of course, NATO as we had known it had changed many years ago.[22]

The foreign policy spokesman of The Netherlands Labour Party, Maarten van Traa, saw the NATO-Russia Council as a possibly valuable body for transparency, but asked the key questions. "Who decides what issues are subject to consultation or co-decision? Will not Russia treat any invitation to enlarge NATO—and perhaps even all peace support operations—as a subject for at least *de facto* 'joint decisionmaking'"?[23]

And what was meant in Helsinki by the "profound transformation" of NATO? Alliance planners could not simply ignore a Russia in transition, but it was surely curious that the then *current* NATO "Strategic Concept," adopted in November 1991, stated that "Soviet [sic] military capability and build-up potential, including the nuclear dimension, still constitute the most significant factor of which the Alliance has to take into account in maintaining the strategic balance in Europe."[24] These anachronistic references to the USSR hardly could be expected to avoid what Ambassador von Moltke had termed "profound" Russian misunderstanding about NATO, and perhaps even contributed to the assertion made by Russian prime minister Viktor Chernomyrdin that NATO must "undertake the obligation to be transformed from a military organization into a political one which would not treat Russia as its main enemy and source of threat."[25] Yet, by such implausible statements Russia sought to eliminate Article 5 of the Washington Treaty, whereas the deputy chairman of the Duma Defense Committee, Lt. General Mikhail Surkov, volunteered the idea that the United States should pledge to withdraw from Europe by 2000.[26] The idea of alliance "transformation," even as it was well under way in terms of enlarging both NATO's missions to peace support and its membership, harked back to the discussions surrounding German unification noted in chapter 1, but adaptation of the Strategic Concept was already foreseen by the alliance.

Secretary of State Albright, in testimony on April 23 before the Senate Armed Services Committee, described Kissinger's analysis as "wrong." She assured senators that the sanctity of NAC decisions would not be affected: "Russia will have a voice, but not a veto." She also said that "both sides will retain complete freedom of action when we cannot agree," and enlargement will not be delayed, whether or not a Russia-NATO accord is in hand.[27] The new secretary of defense, former Republican senator William Cohen of Maine, also stated that the alliance would reserve its right to reassess its policy on the "three noes" regarding nuclear posture should the security environment change. He said that NATO would not accept any limit on its infrastructure and fully intended to modernize reception and reinforcement facilities in new member states[28]—even though Ambassador von Moltke had described the "three noes" and the nonstationing of substantial combat forces as intended "to address Russian concerns."[29] Nevertheless, the Helsinki language was sufficiently vague, whether or not left as a political declaration or a treaty, to lend itself to different interpretations and lead to misunderstandings on the part of both parties. Thus, the European Security Act of 1997, passed in the House on June 11, cautioned that in seeking to demonstrate to Russia NATO's defensive and security-enhancing intentions, fundamental U.S. security interests and alliance effectiveness must not be jeopardized. Specifically, the act stated:

• No commitments should be required, including in the CFE Treaty adaptation negotiations, of new NATO members different from those applicable to present members, including nuclear weapons deployments, the stationing of troops and equipment, the construction of defense infrastructure, and the ability of NATO to reinforce;

• No international organization or nonmember state could "review, delay, veto, or otherwise impede" NAC deliberations and decisions or impede relations between NATO and nonmember states "by, for example, recognizing spheres of influence in Europe," also applied to CFE Treaty adaptation; and

• Russia should offer *reciprocity* in the NATO-Russia document by demarcating all of its borders, stationing forces only with the consent of the state hosting those forces (compare the prolonged delay in Russian withdrawal from Moldova), and reducing nuclear and conventional forces in Kaliningrad.

At the time I wrote: "Should perpetually singling out Russia invite third-party mischief in NATO business, it would raise real concerns about appeasement."[30] After six rounds of negotiations, on May 13–14, the "Founding Act on Mutual Relations, Cooperation and Security between NATO and the Russian Federation" was finalized in Moscow and referred for final approval to governments, following Yeltsin's phone conversations with Western leaders on May 12.

The preamble confirms the Helsinki formula that the act is based on "an enduring political commitment undertaken at the highest political level," although Russia tried up to the last minute to reverse the common position that the document would not be legally binding. It declares, rather anachronistically, that "NATO and Russia do not consider each other as adversaries." It records that NATO would examine its strategic concept "to ensure that it is fully consistent with Europe's new security situation and challenges." NATO will "continue to expand its political functions," take on new missions of peacekeeping and crisis management in support of the UN and the OSCE, and develop the European Security and Defense Identity. Thus, Russia could argue that NATO's "transformation" had taken a step forward.

The second part, "Principles," repeats various OSCE commitments and pledges to "strengthen" the OSCE without elaboration. It includes as a shared principle the inherent right of all states "to choose the means to ensure their own security." Although this falls short of the explicit OSCE commitment that states have the right to belong to treaties of alliance, Russia and NATO did pledge to observe all OSCE commitments.

The third section, "Mechanism for Consultation and Cooperation," creates the "NATO-Russia Joint Permanent Council." It reaffirms the Helsinki formula of creating a mechanism for "consultations, coordination, and, to the maximum extent possible, where appropriate, for joint decisions and joint action with respect to security issues of common concern." What is new is the qualifying language that "consultations will not extend to internal matters of either NATO, NATO member states or Russia" (although Russia could easily argue that NATO enlargement is not solely an internal NATO matter but an issue of "common concern"). The act's provisions "do not provide NATO or Russia, at any stage, with a right of veto over the actions of the other nor do they infringe upon or restrict the rights of NATO or Russia to independent decisionmaking and action." Senator Lugar and others made proposals that those states invited to join NATO should have the same privileges as those offered Russia and even fully participate or have observer status in the Permanent Joint Council.

The council will be jointly chaired by the NATO secretary general, a representative of a NATO nation on a rotating basis, and a representative of Russia and its agenda jointly adopted. The act institutionalizes the practice of biannual meetings of defense and foreign ministers, but adds the possibilities for summits and a degree of predictability by setting a schedule for monthly meetings at the ambassadorial level. To support the work of the council, NATO and Russia will establish "necessary administrative structures," implying a dedicated secretariat (which NACC or PFP initially did not have) as well as new office space at NATO. The inaugural meeting would be held no later than four months after the signature of the act.

Russia agreed to establish a mission to NATO, but NATO "retains the possibility of establishing an appropriate presence in Moscow, the modalities of which remain to be determined." This statement reflected uncertainty within NATO about the usefulness of such a presence in Moscow, which seems politically shortsighted in light of the psychological value a formal alliance presence can have—beyond the existing NATO Information Officer system, which relies on NATO nation embassies.

The parties also "encourage expanded dialogue and cooperation" between the NAA and the Russian Federal Assembly, recognizing "the importance of deepening contacts between the legislative bodies of the participating states to this Act."

The fourth section defines "Areas for Consultation and Cooperation." These are more or less identical to the vast array of topics that had been available through NACC or the new EAPC, ranging from the very broad "issues of common interest related to security and stability" to the more specific areas of nonproliferation, peacekeeping, joint operations, and combating terrorism and drug trafficking. "Possible cooperation in theater missile defense" was also listed, as it was in the European Security Act of 1997. In an effort to improve public understanding, it was also agreed to establish a NATO documentation center or information office in Moscow.

The final section, "Military Dimension," saw NATO reiterate—rather than Russia and NATO agree on—the "three noes" on nuclear weapons and its position on the nonstationing of substantial combat forces. However, it "clarifies" these unilateral positions (both of which are subject to change upon NATO decision) to state that NATO does not intend to establish nuclear weapon storage sites or adapt old nuclear storage sites on the territory of new members. Apart from the nuclear storage question, however, NATO states that it will rely on "adequate infrastructure" to implement collective defense, peacekeeping, and exercises. As in Helsinki, the parties pledge to seek an early framework agreement setting forth the basic elements of an adapted CFE Treaty, but Russia was unsuccessful in its attempts to secure any specific NATO pledges on future CFE provisions. In an important future direction, the parties will promote expanded cooperation between military establishments by exchanging reciprocal military liaison missions and exploring a concept for joint NATO-Russia peacekeeping operations.

The result, according to President Clinton on May 14, was that "Russia will work closely with NATO, but not work within NATO, giving Russia a voice in, but not a veto over NATO's business." The search since 1994 for a formalized relationship— what the act termed "a strong, stable, enduring and equal partnership"—had at last come to closure. Yet, it was also, as Yeltsin's spokesman declared, "only the beginning of the struggle in interpreting this agreement,"[31] and on May 19 Yeltsin informed Russian parliamentarians—to whom the act would be submitted for approval—that should the former Soviet republics be invited to join NATO, Russia would have to "reconsider" its relationship with the alliance.

Not surprisingly, Russian Communist leader Zyuganov immediately described the act as a "treaty on complete capitulation."[32]

Ukraine

On May 7, 1997, the first-ever official NATO Information and Documentation Center opened in Kyiv. It differs from the NATO information officer arrangement in Moscow, which relies on NATO nation embassies. The center is headed by a NATO information official; its operating costs are covered by NATO, and its office space at an international relations institute is provided by Ukraine. In line with a joint NATO-Ukraine press statement of September 14, 1995, calling for an "enhanced relationship" both within and outside the PFP that includes more NATO information activity in Ukraine, the center is intended to foster transparency about the alliance, although Ukrainian authorities equally emphasize its helping to bring Ukrainian perspectives on European security to the alliance.

Also on May 7, NATO presented its own proposal for a NATO-Ukraine document. Ukraine had submitted its texts as long ago as the fall of 1996, but its effort to secure a legally binding treaty, a NATO-Ukraine council, and security guarantees had contributed to delay, as had its reluctance to accept an adaptation of the CFE Treaty, agreed to in May 1996, allowing Russia more flexibility in deploying equipment in the Caucasus and Northern Europe. Ukrainian authorities and parliamentarians claimed the alliance was simply affording Kyiv less attention than Russia and had as a result not paid sufficient attention to negotiating the NATO-Ukraine text. Upon opening the information center, Secretary General Solana implicitly confirmed as much by referring to Russia's "special importance." Even so, Russian presidential advisor Sergei Karaganov interpreted the May 7 event in Kyiv as NATO's "trying to play on contradictions between Russia and Ukraine."[33]

As noted, the NATO-Russia and NATO-Ukraine relationships raise different issues. A Ukrainian public opinion poll of May 6 showed 37 percent in favor of joining NATO, with 28 percent opposed but 34 percent undecided.[34] Ukraine is a nation where fears of division in the Ukrainian and Russian-speaking (22 percent) populations are never far below the surface, and Ukrainian-Russian differences on the future of Russia's naval presence headquartered in Crimea have been a source of mistrust. Ukrainian officials have suggested that its neutral status may not endure forever, and the question may eventually arise whether Ukraine will be regarded as "any other European state" able to enter NATO under Article 10 of the Washington Treaty. Until then, it will be important that NATO not be perceived as regarding Ukraine as a mere ex-Soviet appendage or junior brother of Russia.

Affordable Stability?

In the post-Cold War era of sharply declining defense budgets, the costs of NATO enlargement and how and whether they could be equitably shared by new and future members will prove an important factor as enlargement proceeds. This factor could easily aggravate the long-standing transatlantic burden-sharing debate about whether allies are equitably sharing responsibility in terms of costs, force contributions, and willingness to act. The ratification debate would coincide with the drawdown in the spring of 1998 of the SFOR mission; the United States had declared it would withdraw U.S. forces by June of that year, but European allies would seek a follow-on presence, if required, to forestall an unraveling of the fragile Dayton agreement. All these factors could lead to, as Senator Lugar suggests, "a monumental debate."[35]

Although the March 1996 CBO study had been based, as noted, on the worst-case scenario of a new Russian threat requiring active defenses, the presidential *Report to the Congress on the Enlargement of NATO: Rationale, Benefits, Costs and Implications* of February 24, 1997, which had been requested by Congress, was premised on a continuing benign security environment. In that no-threat environment, total enlargement costs for an "initial, limited group of states" (said to comprise four states) were estimated at $27 billion to $35 billion spread out over the years 1997 to 2009—$13 billion for new members to restructure and $22 billion for NATO regional reinforcement and commonly funded direct enlargement costs (command and control, logistics). Within this range, the United States was expected to pay up to $200 million per year—*one-tenth of 1 percent of the U.S. $260 billion defense budget*, and the other allies would pay *less than 1 percent of their annual defense budgets*. These figures were only estimates, and the report assumed procurement of refurbished *Western* aircraft. But the report indicated all the same that enlargement would not prove financially impossible—if current and future members were willing to make the financial commitments to achieve "initial capability" by 2001 (language training, enhancements to command, control, communications, intelligence, reinforcement reception capability, and air traffic control) and "mature capability" by 2009 (further improvements in the aforementioned categories plus air defense, airfield, road, rail, port, stating area, and fuel storage and distribution improvements)—a roadmap that was the first detailed indication to partners, albeit coming from the Pentagon and not the alliance as a whole, of how to join. The report also assumed that present non-U.S. allies would contribute the bulk of enhanced NATO reinforcement capabilities as part of enlargement (the United States already met these deployability standards) and would thus improve the overall European contribution to reinforcement in any direction.

According to the chairman of the NATO Military Committee, General Klaus Naumann, NATO membership did not mean a "free ride on defense," nor did it imply that new members would have to embark on an ambitious armaments program. The priority—inter-operability for collective defense and NATO's new missions—indicated "no need for NATO to ask future members to standardize equipment by immediate procurement of modern Western material." General Naumann predicted that "the level of additional expenditures for both old and new members could be rather modest and should be seen as an effort stretched over some ten years or so. Moreover, the overall cost for defense will, in the long run, be cheaper than the cost the countries would have to bear if they went for a national defense." (Bypassing the need for a national defense had always been, of course, a key alliance political objective.) General Naumann also noted that none of the present allies contributed more than 0.5 percent of their national defense expenditure to NATO's common budget ($1.7 billion in 1997), of which national shares varied from 0.04 to 40 percent.[36]

Closing Arguments

In pointed contrast to the mere speculation in 1993 about enlargement that accompanied the run-up to the PFP, Secretary of State Albright informed the Senate Armed Services Committee on April 23, 1997, that alliance enlargement was "among the most significant foreign and defense policy issues of our time." A wider NATO, the secretary assured, was essential. It would

- prevent future conflict in Europe, because alliances reduce the likelihood of threat or force being used against them;
- defend democracy and integration;
- "right the wrongs of the past"; and
- strengthen NATO by adding "capable new Allies."

The secretary added that "if there were a major threat to the peace and security of this region, there is a high likelihood that we would decide to act, whether NATO enlarges or not. The point of NATO enlargement is to deter such a threat from ever arising."[37] Of course, this was obvious for many years to me and many colleagues in the Senate and the House and in the NAA, but on the enlargement issue the second Clinton administration found a tighter, mutual bipartisan approach in the White House and on Capitol Hill than ever before (recall the earlier debate over PFP).

Elsewhere, however, it was not just Russia that maintained a "negative" view of enlargement. Even as late as early 1997, an initial skeptic of NATO (Europe's post-World War II anxieties were "a little silly," and the USSR should not be "humiliated")[38] and subsequent author of the "containment" strategy, Professor George F. Kennan, asserted the following in a widely cited editorial in the *New York Times* (a newspaper whose editorial position did not favor NATO enlargement):

• The "impression" that NATO would enlarge was a surprise coinciding with the 1996 U.S. presidential elections and had not allowed for public discussion;
• Enlargement "would be the most fateful error of American policy in the entire post-Cold War era" because it would inflame "the nationalistic, anti-Western and militaristic tendencies in Russian opinion," and the Russian public and government would react by looking outside the West for "guarantees of a secure and peaceful future for themselves";
• "East-West relations" should not be centered on who is allied against whom; and
• There was a "total lack of necessity for this move."[39] Another view regarding the last point about "necessity" is that those nations who "needed it [NATO] would not get it; those that got it would not need it."[40]

Senator Joseph Biden, ranking Democratic member of the Senate Foreign Relations Committee from Delaware, stated that "expansion of NATO has great political appeal here in this country [United States] because of the various ethnic American groups," but that he was not sure how the U.S. public would react "if and when they found out the cost of that expansion, maybe at the expense of the domestic programs in their view or other factors."[41] He also questioned whether the candidate nations were willing to make the necessary financial sacrifices, having stated in Hungary that he expected "recognition in your budget that you are willing to spend more money for defense."[42] (Some Hungarian officials tended to give the impression that the ability of other NATO allies to use bases on Hungarian soil was in itself a major contribution, or that procurement decisions could be postponed for several years until new technologies entered present NATO member forces; systems employing present-day technologies could then be sold to them at a discount.)

During the Senate Armed Services Committee hearings on April 23, the State Department and Pentagon were not entirely prepared for some of the reactions. John Warner, Republican senator from Virginia and former U.S. secretary of the navy, compared a possible future NATO engagement in Central and Eastern Europe to the situation in Somalia and asked whether instability would be bred in the region by inviting only some nations. Why, he wondered, should the United States "foot the bill" in the absence of a threat, and why "fix" an alliance that was not "broken"? His questions reflected the so-called NATO "purist" school: the alliance worked relatively well with sixteen allies and enlargement could erode its effectiveness and credibility. Democratic senator Ted Kennedy of Massachusetts argued that "ten times as much effort has been spent on NATO enlargement compared to what seems to be a much more deadly, clear and present danger"—that is, the surety of nuclear weapons and materiel in Russia. Democratic senator Jeff Bingaman of New Mexico suggested that enlargement would not eliminate but would merely move a dividing line and that the instabilities NATO membership could ameliorate applied equally to Russia. Republican senator Dirk Kempthorne of Idaho wondered whether NATO, as a military alliance, was the right vehicle to foster democracy, and whether enlargement would create "a situation [where] we will have to defend these new members, because we have inadvertently caused this sense of isolation with the very country . . . that still has the capability of destroying the United States of America." Senator John Glenn, Democrat of Ohio, speculated on whether NATO outreach would isolate China and raised the possibility of observer status for China in NATO.[43]

The WEU assembly recommended that states should be free to join the WEU but not NATO so as to provide "imaginative solutions for security arrangements involving European non-NATO members." This would accommodate, it argued, both neutral EU members and NATO aspirants such as the three Baltic states of Estonia, Latvia, and Lithuania.[44] The deputy speaker of the Russian Duma Mikhail Yuriev argued that he was not against WEU enlargement even though it was a military organization, for "the great difference with NATO is that the United States is not part of it, and we can tolerate that a union of this type should be at our doorstep were it not to include that country."[45]

These contentions were easily set aside by my House colleague, Representative Porter Goss of Florida, Republican chairman of the House Intelligence Committee, and Danish parliamentarian Annette Just (Progress Party):[46]

• NATO enlargement was "no surprise" and simply recognized the status quo established in 1989;
• Neither NATO nor a wider NATO required the "necessity" of a threat (as the presidential report on enlargement explicitly took into account), but risks did exist that could best be addressed collectively, as in IFOR/SFOR;
• Alliance values were not the exclusive property of the sixteen allies;
• Opinion polls showed a larger proportion of Russian public opinion—44 percent—indifferent to NATO enlargement than opposed—41 percent;[47] 70 percent viewed Lithuania's joining NATO as Lithuania's business with only 22 percent believing its joining would worsen relations with Russia.[48] In March, Aleksandr Lebed, as he launched his campaign for the presidency, actually *welcomed* NATO enlargement as helping to bring Central Europe into the center of a prosperous Europe;
• The keen interest Russia showed in joining the Group-of-Seven industrialized nations, and even the EU, strongly suggested that a continued orientation toward the West, of which NATO is an indispensable part, would have an essential role if, as jointly stated at the U.S.-Russia Helsinki summit, Russia were to become "a prosperous market-oriented country and a fully-fledged member of leading organizations which will determine economic and trade relations in the 21st century";
• It was "grossly misleading" to suggest that NATO enlargement was primarily a function of appeals by Americans with roots in Central and Eastern Europe. An October 1996 poll found 62 percent of U.S. public opinion in favor of enlarging NATO—although only 7 percent of Americans have their origins in the region toward which the alliance would enlarge; 65 percent of American opinion would even favor admitting Russia if it proves stable and peaceful over time. Although support diminished to 46 percent if the additional cost to the U.S. defense budget were

$1 billion a year,[49] this estimate was not even close to what the Clinton administration envisaged.

• Russian defense minister Igor Rodionov cautioned in February that there was a risk of losing the Russian armed forces as a combat-capable entity by the year 2003 (an outcome he claimed that NATO would welcome).[50] Some commanders have had no experience with live exercises; fatal "hazing" or suicide of soldiers occurs daily; less than 2 percent of the Russian defense budget goes to procurement; arms transfers to regions of tension or to terrorist states go unaccounted for; the early warning system is obsolescent; corruption and theft are reported throughout the ranks (some 160,000 small arms were stolen in 1996); and even elite units start strikes to compel the payment of wages. If there truly were such a risk, then it was precisely the PFP, with its inherent benefits for all partners, and the closest possible relationship with NATO and NATO nations that could help Russia rationalize its 1.7 million forces and move into a streamlined, disciplined, even all-volunteer, service in the next century. It was not, as Senator Kennedy suggested, that there was some connection or choice between nuclear safety and NATO enlargement, but rather that NATO nations and the PFP could be part of Russian military reform *if* the mentality in Russia that "participation in PFP programs is not attractive for career-minded officers" could be abandoned.[51]

• The European Security and Defense Identity (ESDI), to be credible, with the WEU serving as the defense arm of the European Union, must be developed *within* NATO and, hence, *with* the United States, as was explicitly affirmed by all NATO members in Berlin on June 3, 1996. More broadly, as Secretary General Solana put it:

> Our idea of Europe is not limited to economic well-being but reflects a broader set of values—values which unite us with the democracies of North America. After 1945, when Western Europe was given another chance, it was given an Atlantic chance. The same chance, not a lesser imitation of it, should now be given to the new democracies to NATO's East.[52]

Nevertheless, with the Madrid summit only weeks away, Jeremy Rosner, appointed in March to a new post—special adviser to the president and secretary of state for NATO enlargement ratification—concluded that "there are lots of people both in Congress and in the public who have not given much thought to this issue." "We have a lot of work to do," he stated, which includes the fostering of coordination between the NATO national legislatures.[53] Whatever was to be decided in Madrid would have to be implemented.

10

From Paris to Madrid: The Defining Moment

In the remaining weeks before the Madrid summit, all of the essential NATO outreach decisions fell into place. At 10:30 A.M. on May 27, the NATO-Russia Founding Act was signed at the Elysée Palace in Paris. President Clinton, echoing his curious remarks in Helsinki that NATO was but a "mirror image" of the Warsaw Pact, now declared that NATO "will be an Alliance no longer directed against a hostile bloc of nations but instead designed to advance the security of every democracy in Europe"— as if collective defense was a confrontational posture, or as if the Warsaw Pact had not vanished seven years ago. French president Jacques Chirac added no small panache by pronouncing that "the Paris Accord does not shift the divisions created in Yalta, it does away with them once and for all"—although some French deputies saw a U.S.-Russian "condominium" at the expense of the Europeans, themselves divided about "Europe" as an entity on the international stage. President Yeltsin predictably restated opposition to NATO enlargement and wrongly claimed that the act contained "an obligation to non-deploy on a permanent basis combat forces of NATO near Russia" (the act only restates a unilateral NATO policy dependent on the security environment), but announced that the act had taken into account Russian interests and could initiate "a new phase in the life of Europe, peaceful Europe."[1]

Some of this rhetoric may have seemed astonishingly anachronistic, as if before May 24, 1997, there had been a risk of war.

Nevertheless, the key, as Secretary General Solana put it, was now "to give life" to the act.

Polish foreign minister Dariusz Rosati described the signing as meaning that Russia had "in practice and in fact" agreed to NATO enlargement. Still, Russian legislators at the May 27–June 1 NAA spring session in Luxembourg asked "against whom" (*protif kavo*) NATO was enlarging and asserted that because the anti-NATO grouping in the Duma comprised a majority of deputies, the Russian people were *ipso facto* against enlargement. Russian academics themselves acknowledged, however, that "there is no national consensus on the question of NATO enlargement to the East which Russian politicians like to talk about,"[2] and presumably the powerful Russian industrial and financial elites would hardly have an interest in any new confrontation with the West. Yet, even Lebed, who had earlier made supportive remarks about NATO enlargement, now described the act as a "capitulation" (the same word used by Zyuganov) to the "unbridled expansion of the West" and stated that if elected president he would not honor it.[3] Would it be business as usual if the Russian government and Duma formally were to resolve that the viability of the act depended *inter alia* on the permanent exclusion from NATO of former Soviet republics?

Another complication was that President Yeltsin in Paris described working out an OSCE European security "charter" as an "obligation" set down in the act. This was yet another misstatement, for neither the act nor the 1996 OSCE Lisbon document *required* adoption of such a charter; participating states would merely "consider" the possibility. The "charter" notion, according to a February 14, 1997, Russian proposal, would include security guarantees pledged by all 54 participating states, thus possibly raising a conflict with the Washington Treaty (the treaty is commonly understood to require defense only against an attack emanating from the outside and not from within). The charter proposal would also have the OSCE coordinate other organizations, which some viewed as an extension of the Soviet effort to elevate the CSCE during German unification and, now through the charter, to obtain the long-sought *droit de regard* over the alliance, or at least diminish the role of NATO as an organization for European security.

In addition, no doubt Russia would continue to measure NATO's "transformation," noted in the act, by progress toward constricting or even reducing U.S. influence in Europe. This approach is consistent with traditional Soviet diplomacy, albeit framed as supporting "Europeanization."

Hence, it remained to be seen whether, as NATO secretary general Solana stated, the act "puts to rest the notion that NATO and Russia are forever locked in an adversarial posture."[4] Zbigniew Brzezinski observed that should Russia use the Permanent Joint Council for "obtuse and heavy handed" attempts to disrupt internal NATO deliberations, the significance of the council and its recognition of Russia as an equal partner with NATO would be minimized (although the opposite could hold true for "Russia firsters" in the West).[5] Nevertheless, NATO had insisted that it would issue invitations to new members in Madrid whether or not the act had been concluded by then, and it proved that the act was concluded six weeks before. Russia could have chosen to stall and attempt to divide the alliance, with the resulting unpredictable consequences for Madrid (earlier in the year General Naumann, NATO Military Committee chairman, could only state "enlargement decisions, *in my view*, must not be delayed"), but in the end this outcome was avoided.[6]

That afternoon, at 3:15, the NAC approved by silence procedure the "Charter on a Distinctive Partnership" between NATO and Ukraine. It established a "commission" to review implementation of the cooperation menu twice a year—not, as Ukraine sought, a special consultative mechanism akin to the Russia-NATO Permanent Joint Council (likewise Kyiv preferred a "special partnership"). No sought-after security guarantees were provided, as they apply only to NATO members. The following day, Russia and Ukraine agreed a friendship treaty recognizing their borders and resolving, in principle, the prolonged Russia-Ukraine dispute over the former Soviet Black Sea Fleet with an agreement to continue basing Russian ships in Sevastopol but to do so jointly with Ukrainian ships (Russia sought exclusive use) and on a lease basis for 20 years (set off against the multibillion-dollar Ukrainian energy debt to Russia)—an outcome perceived as formal Russian recognition of Ukraine as an independent nation. (In early May, Belarus, which maintained its partner status despite its regression from a democratic path, also raised the issue of a special agreement with NATO.)

Meeting in Sintra, Portugal, on May 29, the NAC reviewed a report on the adaptation of alliance structures to integrate new members, country assessments prepared by the NATO Military Committee, and a plan and timetable for the accession talks. A cost study called for in December 1996 had still not been agreed, with U.S. diplomats criticizing an undetailed, "inadequate and unprofessional approach" and a European proclivity to "lowball" costs. The ministers recommended that in Madrid the NATO heads of state issue an "explicit" commitment that the alliance would remain open to the accession of any other European state and give "substance" to this commitment. Five countries—the Czech Republic, Hungary, Poland, Romania, and Slovenia—were now supported by nine NATO members as well as reportedly by Secretary General Solana, although his role was to act as a "catalyst for consensus": Belgium, Canada, France, Greece, Italy, Luxembourg, Portugal, Spain, and Turkey. The United States, Iceland, and possibly the United Kingdom favored only the first three candidates, whereas Denmark, Germany, the Netherlands, and Norway were undecided.[7]

Among the factors still being discussed, even at this late stage, were NATO's ability to absorb several members (five nations would nearly *triple* present membership), avoiding any impression that the choice of several countries might imply a "one-shot" enlargement, assuring prompt ratification, and providing geographical balance in the first wave. Secretary General Solana sought a final list of invitees by the third week of June, after which he planned to go to candidate nations to inform them of the decisions so that Madrid would bear no surprises.

Of course, as Poland was creating a peacekeeping battalion with Lithuania (and Lithuania with Estonia and Latvia), and Hungary was establishing a like battalion with Romania plus a joint brigade with Italy and Slovenia, both current and future early members could help bring the follow-on wave nearer. All the same, meeting in Otepää, Estonia, on May 26, the three Baltic presidents appealed for explicit support in Madrid for Baltic membership, regular review of the enlargement process, and a commitment to subsequent invitations no later than when the first new members joined (the goal being 1999).

It was also in Sintra on May 30 that the 44-nation EAPC was inaugurated, replacing the NACC and subsuming the PFP. Russian first deputy foreign minister Igor Ivanov made a point to describe the EAPC as "not a school to prepare for future NATO members" but one to prepare for addressing new challenges.

Whether in its political dimension the EAPC proves of greater interest than the NACC is an open question, but the enhanced PFP, by narrowing the difference between ally and partner, and bilateral assistance such as U.S. foreign military financing grants can help interested partners improve their prospects for NATO membership.

On May 29, the NATO-Ukraine charter was initialed for signing in Madrid. Thus, by the end of May, the NATO-Russia, NATO-Ukraine, and enhanced PFP were decided—even as decisions on NATO's *internal* reform concerning new command structures and the European role lagged well behind. What remained was a decision on the first new members and a convincing response to those fearing that the NATO door would be closed after Madrid—an outcome Russia would certainly lobby NATO member governments to condone. German foreign minister Kinkel suggested that this could be achieved by restating the Copenhagen NAC language from 1991 about how the security of European democracies was of direct and material concern to the alliance and how it should be NATO policy to "support" nations that share the alliance's values—candidates for EU membership "including the Baltic states."[8]

As for these nations' future NATO membership, German junior foreign minister Helmut Schaefer argued in June that, after the first round, thought should be given to an "alternative concept" of security "before more harm is done" in Russia. He cautioned "all those in Germany who, out of jingoism, want to take all sort of countries into the Alliance."[9] In Sintra, Secretary of State Albright informed the NAC that it would be "essential" to establish a new phase of dialogue with partners not invited to join in the first wave and that the Madrid summit could not be treated "as the first and last chance to join NATO."[10] But opinions differed on whether the U.S. preference for only three countries in the first wave was intended to clarify that additional waves would be coming or to prejudge the next wave with Romania and Slovenia only, postponing consideration of the Baltic states indefinitely. In Sintra, Secretary Albright rightly declared that "no European country will be excluded because of where it sits on the map."[11] Yet did the very fact that the secretary felt compelled to state as much suggest anticipated difficulties within the alliance?

In any event, although the formal decision to invite members had been delayed until December 1996, the situation was that NATO members had less than seven weeks to achieve consensus on who the "one or more" countries to be invited would be, and the Clinton administration, to which a number of allies would look to, claimed that it had not reached any decision itself.

If the protocols of accession could be achieved by the end of 1997, then by the spring of 1998 the U.S. Senate would undertake the process of advice and consent to ratification, and many other NATO national parliaments might await a U.S. decision prior to taking their own decisions. Already in July 1996 NATO had requested the 16 nations to provide information on national administrative and legal procedures for ratification (which might not be required in Canada, Norway, or the United Kingdom), with the conclusion being that the whole process could take up to a year.

Yet concerns were expressed by Senator Lugar and others that the administration was not doing enough and working sufficiently with Congress to clarify essential issues prior to the Madrid summit, for the debate about enlargement would be invariably tied to the broader role of the United States in European security, including a future U.S. presence in Bosnia. Representatives Floyd D. Spence and Ronald V. Dellums, respectively the republican chairman and the ranking democratic minority member of the House National Security Committee, argued at the end of May that seven fundamental questions must be answered "to illuminate the purpose, function, structure and membership of an expanded NATO" (notwithstanding that the 1996 NATO Enlargement Facilitation Act clearly declared that "the admission to NATO of emerging democracies in Central and Eastern Europe . . . would contribute to international peace and enhance the security of the region"):

1. What role would the alliance play in U.S. national security strategy beyond collective defense?

2. Are realistic criteria for membership being followed—for example, might insistence on a free market go beyond the requirements for membership?

3. What does Article 5 mean today, given that its ambiguous phrasing requires "such action as [an ally] deems necessary" in the event of an attack on another?

4. Could new security commitments be reconciled with re- duced defense expenditures, and would the nonstationing of nu- clear weapons on new member states mean greater reliance on strategic weapons for nuclear deterrence?

5. Even if the White House cost estimates were accurate, what expenditures can be foreseen beyond the initial enlargement?

6. Because the NATO-Russia act would probably not resolve all issues, might NATO enlargement become a permanent source of tension with Russia in an already complex relationship?

7. Where do the European nations stand on new members, and might new allies import new tensions?[12]

Similarly, on June 25, 20 senators, including chairman of the Foreign Relations Committee Senator Jesse Helms, wrote Presi- dent Clinton raising 10 sets of questions that, they argued, re- quired answers prior to the first admission. Their list began as follows: "What is the military threat that NATO expansion is designed to counter?"[13]

Some of these questions, of course, were not directly related to enlargement or could only be clarified after Madrid; some even implied that NATO as an element of U.S. national security strat- egy required perpetual revalidation. Again, it was Congress itself that had led the way on the enlargement issue. Nevertheless, in an address at West Point on May 31, President Clinton acknowl- edged the need "to have an open, full, national discussion before proceeding" with enlargement, and he cited four reasons for a larger alliance. Enlargement would

- strengthen the ability of the alliance to meet future security challenges, as in Bosnia;
- secure democracy;
- encourage prospective members to resolve their differ- ences by peaceful means; and
- "erase the artificial line in Europe that Stalin drew, and bring Europe together in security"—goals that an enhanced PFP and special NATO relationships with Russia and Ukraine would also help achieve.

The president also rightly stated: "Some say we no longer need NATO because there is no powerful threat to our security now. I say there is no powerful threat in part because NATO is there. And enlargement will help make it stronger.[14]

On June 12 the White House broke silence on the "who" of the first wave. President Clinton announced:

> After careful consideration, I've decided that the United States will support inviting three countries—Poland, Hungary and the Czech Republic—to begin accession talks to join NATO when we meet in Madrid next month.
>
> We have said all along that we would judge aspiring members by their ability to add strength to the Alliance and their readiness to shoulder the obligations of NATO membership. Poland, Hungary and the Czech Republic clearly meet those criteria—and have currently made the greatest strides in military capacity and political and economic reform.
>
> As I have repeatedly emphasized, the first new members should not and will not be the last. We will continue to work with other interested nations, such as Slovenia and Romania, to help them prepare for membership. Other nations are making good progress—and none will be excluded from consideration.[15]

White House spokesman Mike McCurry clarified on June 17 that Slovenia still had to make progress in military terms, Romania in political and economic reform.

Secretary of Defense Cohen explained on June 12 the advantages to this limited opening: it would diminish the problems and costs, take advantage of the consensus on these three states, and underscore that enlargement will continue for countries on the "right path." At the same time, he urged that the alliance clearly commit in Madrid to keeping membership open, continuing the intensified dialogue, and reviewing the progress of additional nations on an ongoing basis. Yet, this three-nation first wave view was immediately contested by several allies, most intensely by Italy and France. Even though the enlargement choice would, as with any NATO decision, require consensus, there was resentment over what some allies saw as a late-in-the-day U.S. effort to impose its view on the alliance.

As events evolved, the final decision had to go right to the summit itself. On the morning of July 8, meeting in the Palacio Municipal de Congresos in Madrid, the NATO leaders affirmed the Sintra consensus on the Czech Republic, Hungary, and Poland (these three countries, plus Cyprus, Estonia, and Slovenia, were then recommended by the European Commission on July 15 for accession negotiations with the EU, albeit with a view to membership in 2004). The goal would be to sign the protocols of accession to the North Atlantic Treaty later that year at the NAC meeting in Brussels on December 16. Senator Roth, attending the summit as NAA president and as chairman of the 28-member Senate NATO Observer Group established by the bipartisan Senate leadership, predicted that the votes would be there in support of these countries, although greater clarity would be required, he signaled, from current and prospective members about bearing their fair share of the burdens, risks, and costs.

Following what U.S. national security adviser Sandy Berger described as a "lively debate" that day, at 5:00 P.M. Secretary General Solana was able to announce the decision on the three countries and the results of the long discussion on how to find a formula that would keep the door open to Slovenia and Romania, as well as to others. Paragraph 8 of the declaration could be read as reflecting a so-called three plus two in two years formula (although "neutral" Austria also could well apply for a second wave, opening the way for a broader congruence of security interests between NATO and EU members). This paragraph left plenty of scope, however, for those preferring caution following the first wave:

- "The Alliance expects to extend further invitations in coming years";
- No European democracy would be excluded from consideration "regardless of their geographic location," which President Clinton stated applied, in his personal view, to Russia;
- Future membership would have to serve the overall political and strategic interests of the alliance and "enhance overall European security and stability";
- NATO would continue its intensified dialogue with all interested nations and keep the process of the "open door" under continual review;
- Great interest was shown in, and account taken of, positive developments "in a number of southeastern European countries, especially Romania and Slovenia";

- The alliance also recognized progress toward stability and cooperation "by the states in the Baltic region which are also aspiring members";
- The "ongoing enlargement process" would be reviewed at the next summit in 1999.

"Today's meeting is a defining moment for NATO," Solana declared. "Madrid will be remembered as the time when North America and Europe came together to shape the course of a new century. . . . Our Alliance will emerge stronger from Madrid and ready to assume all the new tasks we have set for ourselves."

Attending the summit in the absence of President Yeltsin, Russian deputy prime minister Valeri Serov restated opposition to NATO enlargement, but also stated that Russia had "to take into account existing realities" and stood for "cooperation with NATO for the sake of consolidated European security."[16] And to help define those realities, the alliance leaders also directed that the NAC "bring to an early conclusion the concrete analysis of the resource implications of the forthcoming enlargement," expressing confidence that enlargement costs "will be manageable and that the resources necessary to meet those costs will be provided."

Conclusion and Recommendations

Our Alliance has come far on its way to transformation. Its relevance today does not derive from its original and immediate purpose but from what it has become over time. It has evolved into a community of values and destiny, and a forum of political consultation on vital issues of foreign policy and security. It has evolved into an agent of change. It will become the core security organization of a future Euro-Atlantic architecture in which all states, irrespective of their size or geographical location, must enjoy the same freedom, cooperation and security. We must not be satisfied with having won the Cold War. We have to win the future.

—NATO Secretary General Manfred Wörner
October 21, 1991

This inquiry has highlighted in some detail the political and defense issues decision-makers were repeatedly compelled to address regarding the role of the Atlantic Alliance in an era of rapid change. NATO governments, legislatures, and policy elites were challenged, not without difficulty, with adapting NATO to the emerging security environment and rediscovering the original, broad purpose of the Atlantic Community to encourage habits of cooperative security among nations—"a just and lasting peaceful order in Europe accompanied by security guarantees."

On the one hand, the NATO response, which was evolutionary and cautious, is consistent with the historical experience of 16 sovereign nations reaching critical common security decisions by consensus. The impetus for enlargement came largely from the new democracies of Central Europe. The prolonged debate within NATO governments about whether enlargement should turn on the appearance of a new threat from Russia—as if NATO required a clear and present danger or was no more than a purely military organization of convenience only to be held in reserve without actively improving security in the whole of Europe for members and nonmembers alike—clearly evinced the thinking of a bygone era, even if such old thinking in official Russia had to be "taken into account." Responding to the initiative from Central Europe also intersected with the parallel demands of NATO's adaptation to the new Europe in other ways—namely, projecting power beyond the treaty area in support of peace, enhancing the European security and defense identity within the alliance, forging a strategic partnership with Russia, and extending NATO's values and habits of cooperation to all partners.

On the other hand, within only slightly more than two years of the PFP compromise, which prompted congressional and NAA reaction that helped maintain the momentum for enlargement, the Madrid NATO summit decided that the time for NATO's fourth opening had come—not to move a front line eastward, but to enable qualified nations abandoned by democracy for two generations to rejoin the West at the expense of no other state. The alliance took all opportunities to provide Russia reassurance at every stage, as transpired with German unification, although insisting that how new members join NATO would remain NATO business. As we look toward those for whom the NATO door will remain open beyond the Madrid decisions, we have clearly reached the stage, one hopes, where the *only relevant question* will be not *whether* but *to whom* the benefits and responsibilities of NATO membership will be extended—those very same opportunities, as the Washington Treaty requires, that are offered to *any other* European state. The "open door" cannot be allowed to become an eternal delaying tactic or a process contingent on de facto consent from Russia, and no state can be excluded solely because of geography.

It is a vitally important lesson, as we consider the future of transatlantic and European security, that none of these changes would have occurred without active and persistent U.S. leadership. Just as Washington proved the key to ensuring that security for all of Europe would best be served by a Germany united within the integrative NATO community rather than adrift in an undefined grey zone in the middle of Europe, so too did U.S. leadership prove decisive in applying the same integrative logic to other nations. A continued bipartisan approach will remain instrumental as the process of NATO's transformation and opening proceeds, just as it did when the United States decided in 1949 to engage in this "entangling alliance" that proved one of the greatest of foreign policy success stories.

As a result, the path toward righting a historical wrong has been opened. Yet, the NATO decisions in Madrid are only the beginning of a process and will require close scrutiny. I see five priorities:

• First, the alliance must have an agreed basis for contributing to the financing of a wider NATO.

• Second, if a wider alliance is to provide a broader basis for responsibility-sharing, the U.S. administration must regularly report on the progress new members and those to follow are making toward meeting criteria, what specific and coordinated assistance will be required, and how all allies are contributing to resourcing enlargement and adapting their forces for power projection and in meeting the threat of the proliferation of weapons of mass destruction—a challenge in which Russia should be encouraged to join as an early test of the NATO-Russia Act. We have been through the burdensharing debate many times, and cannot allow disagreement and the lack of meaningful information to detract from our commitment to the new democracies, who of course will have to assume their fair share as well.

• Third, and although this issue is divorced from enlargement, the recurring debates regarding NATO's role must be better managed. Clearly, NATO's decisions in 1992 to offer its assets in support of peace operations under UN mandate or OSCE authority are not settled, and the debate about the post-SFOR situation will coincide with that on ratification of enlargement. The noble vision of the Madrid Declaration notwithstanding, the new Alliance Strategic Concept for the Twenty-first Century must be the subject of transparent debate and legislative scrutiny before it is adopted if it is to serve as a meaningful compass for the future.

• Fourth, it is self-evident that there can be no security in Europe and NATO as a whole without Russia. Although we must be clear that enlargement of the alliance is NATO business, we must work to make the NATO-Russia act a living document. Our goal must be to prevent inertia and a psychology of "doomed to coexistence" from taking the place of real partnership and cooperation.

• Finally, an invigorated alliance can only go forward with the support of public opinion. Legislators from present and future members must intensify their dialogue, with the NAA being the natural forum, and NATO itself must become more transparent.

In the final analysis, a wider alliance is but a means to the end of building confidence and security toward which all of NATO's directions are aimed. In an era of profound transformation in transatlantic and European security, there can be no guarantees that the values and strategic outlook of the alliance can form the foundation for all of Europe. Nevertheless, we do know that the NATO experience has much to offer as we return to the original broad ambition of NATO and embrace a wider community of free peoples.

Appendix A

Madrid Declaration on Euro-Atlantic Security and Cooperation

Issued by the Heads of State and Government
[excerpts]

1. We, the Heads of State and Government of the member countries of the North Atlantic Alliance, have come together in Madrid to give shape to the new NATO as we move towards the 21st century. Substantial progress has been achieved in the internal adaptation of the Alliance. As a significant step in the evolutionary process of opening the Alliance, we have invited three countries to begin accession talks. We have substantially strengthened our relationship with Partners through the new Euro-Atlantic Partnership Council and enhancement of the Partnership for Peace. The signature on 27th May of the NATO-Russia Founding Act and the Charter we will sign tomorrow with Ukraine bear witness to our commitment to an undivided Europe. We are also enhancing our Mediterranean dialogue. Our aim is to reinforce peace and stability in the Euro-Atlantic area.

A new Europe is emerging, a Europe of greater integration and cooperation. An inclusive European security architecture is evolving to which we are contributing, along with other European organisations. Our Alliance will continue to be a driving force in this process.

2. We are moving towards the realisation of our vision of a just and lasting order of peace for Europe as a whole, based on human rights, freedom and democracy. In looking forward to the 50th anniversary of the North Atlantic Treaty, we reaffirm our commitment to a strong, dynamic partnership between the European and North American Allies, which has been, and will continue to be, the bedrock of the Alliance and of a free and prosperous Europe. The vitality of the transatlantic link will benefit from the development of a true, balanced partnership in which Europe is taking on greater responsibility. In this spirit, we are building a European Security and Defence Identity within NATO. The Alliance and the European Union share common strategic interests. We welcome the agreements reached at the European Council in Amster-

dam. NATO will remain the essential forum for consultation among its members and the venue for agreement on policies bearing on the security and defence commitments of Allies under the Washington Treaty.

3. While maintaining our core function of collective defence, we have adapted our political and military structures to improve our ability to meet the new challenges of regional crisis and conflict management. NATO's continued contribution to peace in Bosnia and Herzegovina, and the unprecedented scale of cooperation with other countries and international organisations there, reflect the cooperative approach which is key to building our common security. A new NATO is developing: a new NATO for a new and undivided Europe.

4. The security of NATO's members is inseparably linked to that of the whole of Europe. Improving the security and stability environment for nations in the Euro-Atlantic area where peace is fragile and instability currently prevails remains a major Alliance interest. The consolidation of democratic and free societies on the entire continent, in accordance with OSCE principles, is therefore of direct and material concern to the Alliance. NATO's policy is to build effective cooperation through its outreach activities, including the Euro-Atlantic Partnership Council, with free nations which share the values of the Alliance, including members of the European Union as well as candidates for EU membership.

5. At our last meeting in Brussels, we said that we would expect and would welcome the accession of new members, as part of an evolutionary process, taking into account political and security developments in the whole of Europe. Twelve European countries have so far requested to join the Alliance. We welcome the aspirations and efforts of these nations. The time has come to start a new phase of this process. The Study on NATO Enlargement—which stated, inter alia, that NATO's military effectiveness should be sustained as the Alliance enlarges—the results of the intensified dialogue with interested Partners, and the analyses of relevant factors associated with the admission of new members have provided a basis on which to assess the current state of preparations of the twelve countries aspiring to Alliance membership.

6. Today, we invite the Czech Republic, Hungary and Poland to begin accession talks with NATO. Our goal is to sign the Protocol of Accession at the time of the Ministerial meetings in December 1997 and to see the ratification process completed in time for membership to become effective by the 50th anniversary of the Washington Treaty in April 1999. During the period leading to accession, the Alliance will involve invited countries, to the greatest extent possible and where appropriate, in Alliance activities, to ensure that they are best prepared to undertake the responsibilities and obligations of membership in an enlarged Alliance. We direct the Council in Permanent Session to develop appropriate arrangements for this purpose.

7. Admitting new members will entail resource implications for the Alliance. It will involve the Alliance providing the resources which enlargement will necessarily require. We direct the Council in Permanent Session to bring to an early conclusion the concrete analysis of the resource implications of the forthcoming enlargement, drawing on the continuing work on military implications. We are confident that, in line with the security environment of the Europe of today, Alliance costs associated with the integration of new members will be manageable and that the resources necessary to meet those costs will be provided.

8. We reaffirm that NATO remains open to new members under Article 10 of the North Atlantic Treaty. The Alliance will continue to welcome new members in a position to further the principles of the Treaty and contribute to security in the Euro-Atlantic area. The Alliance expects to extend further invitations in coming years to nations willing and able to assume the responsibilities and obligations of membership, and as NATO determines that the inclusion of these nations would serve the overall political and strategic interests of the Alliance and that the inclusion would enhance overall European security and stability. To give substance to this commitment, NATO will maintain an active relationship with those nations that have expressed an interest in NATO membership as well as those who may wish to seek membership in the future. Those nations that have previously expressed an interest in becoming NATO members but that were not invited to begin accession talks today will remain under consideration for future membership. The considerations set forth in our 1995 Study on NATO Enlargement will continue to apply with regard to future aspirants, regardless of their geographic location. No European democratic country whose admission would fulfil the objectives of the Treaty will be excluded from consideration. Furthermore, in order to enhance overall security and stability in Europe, further steps in the ongoing enlargement process of the Alliance should balance the security concerns of all Allies.

To support this process, we strongly encourage the active participation by aspiring members in the Euro-Atlantic Partnership Council and the Partnership for Peace, which will further deepen their political and military involvement in the work of the Alliance. We also intend to continue the Alliance's intensified dialogues with those nations that aspire to NATO membership or that otherwise wish to pursue a dialogue with NATO on membership questions. To this end, these intensified dialogues will cover the full range of political, military, financial and security issues relating to possible NATO membership, without prejudice to any eventual Alliance decision. They will include meeting within the EAPC as well as periodic meetings with the North Atlantic Council in Permanent Session and the NATO International Staff and with other NATO bodies as appropriate. In keeping with our pledge to maintain an open door to the admission of additional Alliance members in the future, we also direct that NATO Foreign Ministers keep that process under continual review and report to us.

We will review the process at our next meeting in 1999. With regard to the aspiring members, we recognise with great interest and take account of the positive developments towards democracy and the rule of law in a number of southeastern European countries, especially Romania and Slovenia.

The Alliance recognises the need to build greater stability, security and regional cooperation in the countries of southeast Europe, and in promoting their increasing integration into the Euro-Atlantic community. At the same time, we recognise the progress achieved towards greater stability and cooperation by the states in the Baltic region which are also aspiring members. As we look to the future of the Alliance, progress towards these objectives will be important for our overall goal of a free, prosperous and undivided Europe at peace.

9. The establishment of the Euro-Atlantic Partnership Council in Sintra constitutes a new dimension in the relations with our Partners. We look forward to tomorrow's meeting with Heads of State and Government under the aegis of the EAPC.

The EAPC will be an essential element in our common endeavour to enhance security and stability in the Euro-Atlantic region. Building on the successful experience with the North Atlantic Cooperation Council and with Partnership for Peace, it will provide the overarching framework for all aspects of our wide-ranging cooperation and raise it to a qualitatively new level. It will deepen and give more focus to our multilateral political and security-related discussions, enhance the scope and substance of our practical cooperation, and increase transparency and confidence in security matters among all EAPC member states. The expanded political dimension of consultation and cooperation which the EAPC will offer will allow Partners, if they wish, to develop a direct political relationship individually or in smaller groups with the Alliance. The EAPC will increase the scope for consultation and cooperation on regional matters and activities.

10. The Partnership for Peace has become the focal point of our efforts to build new patterns of practical cooperation in the security realm. Without PfP, we would not have been able to put together and deploy so effectively and efficiently the Implementation and Stabilisation Forces in Bosnia and Herzegovina with the participation of so many of our Partners.

We welcome and endorse the decision taken in Sintra to enhance the Partnership for Peace by strengthening the political consultation element, increasing the role Partners play in PfP decision-making and planning, and by making PfP more operational. Partners will, in future, be able to involve themselves more closely in PfP programme issues as well as PfP operations, Partner staff elements will be established at various levels of the military structure of the Alliance, and the Planning and Review Process will become more like the NATO force planning process. On the basis of the principles of inclusiveness and self-differentiation, Partner countries will thus be able to draw closer to the Alliance. We invite all Partner countries to take full advantage of the new possibilities which the enhanced PfP will offer.

With the expanded range of opportunities comes also the need for adequate political and military representation at NATO Headquarters in Brussels. We have therefore created the possibility for Partners to establish diplomatic missions to NATO under the Brussels Agreement which entered into force on 28th March 1997. We invite and encourage Partner countries to take advantage of this opportunity.

Appendix B
Chronology of Principal Events

1988

February 28–March 4   The Subcommittee on Eastern Europe of the NAA Political Committee pays the first ever NAA trip to a Warsaw Pact member state, Hungary

March 21–23   NAA Secretary General Peter Corterier pays first official visit to Moscow

November 14   Hungarian State Secretary for Foreign Affairs Gyula Horn (future prime minister) is first speaker from a Warsaw Pact nation to address the NAA, in Hamburg

1989

July 7–8   Warsaw Pact launches "reform" process

August 25   Hungary allows East Germans to enter West via open border with Austria

November 9–10   Breach of Berlin Wall

December 4   Warsaw Pact denounces 1968 invasion of Czechoslovakia and repudiates "Brezhnev Doctrine"

1990

May 22   Hungary proposes negotiations to leave Warsaw Pact

May 25   USSR suggests unified Germany's political membership in NATO

May 31   USSR suggests German membership in both NATO and Warsaw Pact

June 7	Warsaw Pact declares conditions emerging for "overcoming the bloc model of security and the division of the continent"
July 5–6	NATO extends a "hand of friendship" to "former adversaries"
September 2	Hungarian prime minister Jozsef Antall calls for a Central European Western European Union (WEU)
September 9–12	Treaty on German unification signed in Moscow; Wojciech Lamentowicz of Poland proposes bilateral security treaties between NATO and Czechoslovakia, Poland, and Hungary, which would temporarily remain in the "political" structure of the Warsaw Treaty
November 19	CFE Treaty signed in Paris, NATO and Warsaw Pact member states jointly declare they are "no longer adversaries"
November 28	NAA in London urges Warsaw Pact nation attendance at North Atlantic Council (NAC) and other NATO bodies, extends associate delegate status to Warsaw Pact national parliaments

1991

February 15	In Visegrad, Hungary, the Czech and Slovak Federal Republic, Hungary, and Poland pledge cooperation in seeking total political, economic, and security integration with Europe
February 25	Warsaw Pact military structure dissolved in Budapest
March 11	Bulgarian president Zhelyu Zhelev calls for immediate associate membership in NATO
June 6–7	NATO declares security in Europe as "inseparable," upgrades liaison to participation in NATO-sponsored seminars and courses
June 25	Slovenian and Croation Yugoslav republics declare independence, Serbian republic invades
July 1	Warsaw Pact disbanded in Prague
August 21	NATO suggests differentiation during "coup" attempt against Soviet president Mikhail Gorbachev

October 22	NAA in Madrid opens perspective on new NATO members, urges joint field training exercises among states not parties to same alliance
November 7–8	NATO Rome summit proposes the North Atlantic Cooperation Council (NACC)
December 20	First NACC meeting in Brussels; Russian President Boris Yeltsin raises prospect of Russian membership in NATO
December 21	Soviet Union dissolves

1992

January 10	NATO High Level Working Group addresses post-USSR implications for CFE Treaty
March 10	Extraordinary NACC meeting in Brussels expands to include former Soviet republics; first work plan adopted
April 1	First NATO Group on Defense Matters meeting in Brussels
April 10	First meeting of the NATO Military Committee in Cooperation Session
April 10	WEU Secretary General Willem van Eekelen regrets undifferentiated nature of the NACC and calls for special relationship between WEU and new democracies
June 4	NATO agrees in Oslo to consider CSCE requests for peacekeeping support
June 5	Russia at NACC suggests special responsibility for peacekeeping in the former Soviet Union
June 19	WEU Forum for Consultation established
July 17	CFE Treaty comes into force provisionally
September 2	NAC agrees to provide resources to support UN, CSCE, and EC efforts in former Yugoslavia
September 23	U.S. NATO ambassador Reginald Bartholomew warns against NACC becoming "largely symbolic," urges active cooperation in peacekeeping and crisis management
September 24	Deputy CINCEUR calls for NACC cooperation wing at NATO headquarters, security consultations similar to Article IV of the North Atlantic Treaty, NACC military support for UN and CSCE decisions; predicts NACC can become "the major forum of European security"

December 17	NATO agrees to consider support for implementation of UN Security Council resolutions
December 18	NACC agrees to "consultations . . . leading to cooperation" in peacekeeping

1993

February 11	NACC Ad Hoc Group on Cooperation in Peacekeeping established
April 22	President Clinton begins interagency process considering NATO enlargement
April	Russian-U.S. exercise held in Laptev Sea
May 21	German defense minister Volker Rühe states that admission of the Czech Republic, Hungary, Poland, and Slovakia into NATO and EU is question of "how" and "when," not "if"
June	Baltic, Finnish, Polish, and Swedish forces or observers participate alongside NATO forces in *Baltops* exercise in Baltic Sea
June 17	Charter for American-Russian Partnership and Friendship calls for "credible Euro-Atlantic peacekeeping capability"
August 25	President Yeltsin and Polish president Lech Walesa adopt Joint Declaration stating that "in perspective, a decision of this kind by sovereign Poland [to join NATO] aiming at all-European integration is not contrary to the interests of other states including also Russia"
August 31	Polish prime minister Hanna Suchocka urges NATO to decide on new membership and a timetable, warns of "alternative" means of defense
September 10	NATO secretary general Manfred Wörner calls for "more concrete perspective" to nations seeking NATO membership
September 15	President Yeltsin writes to France, Germany, United Kingdom, and United States arguing that the September 1990 "Treaty on the Final Settlement on Germany" precludes further NATO expansion eastward, urges "warmer" Russia-NATO relations than NATO-Central European relations, offers joint security guarantees to Central Europe

September 23	NATO press spokesman Jamie Shea, in personal comments, calls for NATO accession protocols for Czech Republic, Hungary, and Poland
October 11	NAA in Copenhagen urges timetable and criteria for NATO expansion
October 17	Memorandum from adviser for former Soviet Union Strobe Talbott to secretary of state reportedly rejects NATO "associate" status and membership criteria, as urged by Undersecretary of State Lynn Davis
October 19	White House approves Partnership for Peace (PFP)
October 20–21	U.S. secretary of defense Les Aspin reveals PFP at Defense Planning Committee (DPC) meeting in Travemünde, Germany
October 28	NATO briefed on PFP
December 9	Polish and Czech foreign ministers appeal for NATO summit to signal that NATO "will sooner or later take in as new members the countries of Central-Eastern Europe"

1994

January 8	President Clinton notes absence of allied consensus to expand NATO
January 10–11	NATO summit endorses PFP
January 12	President Clinton declares question of NATO membership as no longer "whether" but "when" and "how"
January 26	Romania first to sign PFP Framework Document
January 27	U.S. Senate urges, by 94 to 3, prompt admission to NATO of qualified nations
February 8	Ukraine first CIS country to sign PFP
February 28–March 1	NATO PFP Briefing Mission informed in Moscow of Russian desire for special relationship in or outside of PFP
April 18	Polish Foreign Ministry official Andrzej Towpik urges NATO security guarantees for PFP Partners; Partnership Coordination Cell activated next to SHAPE
April 25	Poland first to submit Presentation Document
May 9	WEU establishes associate partner status
May 16–20	UK-Polish land exercise in Poland

May 24–25	Russian defense minister Pavel Grachev calls Framework Document inadequate, calls for NATO-Russian "mechanism of consultations on the whole range of European and global security issues"
June 5	U.S. secretary of state Warren Christopher claims PFP carries "certain security guarantees with respect to the territorial integrity of the . . . members of the Partnership"
June 9	NATO affirms alliance enlargement a matter for NATO decision, but describes cooperation with Russia as "key element for security and stability in Europe"
June	NACC Ad Hoc Group on Cooperation in Peacekeeping merges with Political-Military Steering Committee
June 22	Russia signs Framework Document, "Summary of Conclusions"
June 23	Russian foreign minister Andrei Kozyrev proposes CSCE coordinate "division of labor" among CIS, NACC, EU, Council of Europe, NATO, and WEU, proposes CSCE "Executive Committee"
July 1	President Clinton urges NATO to begin examining timetable and standards for new NATO members
July 4	Poland first to conclude IPP
July 5	Russia submits Presentation Document
July 7	President Clinton offers to request $25 million for Polish PFP participation, states NATO expansion does not depend on new threat in Europe
July 14	U.S. Senate urges accession of Visegrad countries
July 19	German defense minister Rühe states Russia will not join NATO
August 31	Russian forces complete withdrawal from Baltic states and Germany
September 12–16	600 troops from 13 NATO and partner nations participate in first PFP exercise, *Cooperative Bridge '94*, in Biedrusko, Poland; Russia only observes
September 27	Republicans propose "National Security Restoration Act" calling for Visegrad mem-

	bership in NATO no later than January 10, 1999
October 7–8	U.S. Congress passes "NATO Participation Act" calling for the transfer of defense equipment to Visegrad and other states to assist in transition to NATO "at an early date"
October	U.S. interagency Working Group on European Security begins considering criteria for NATO enlargement
October 28	NATO secretary general Willy Claes states that next NATO task is to prepare for new NATO members "in a way which enhances European security"
November 14	Former U.S. national security adviser Zbigniew Brzezinski warns of effort to elevate criteria for NATO membership to unachievable extent; NAA urges NATO enlargement to begin within two to three years, timetable and criteria to be provided, and memoranda of understanding on accession protocol to be elaborated by mid-1995
December 1	NATO agrees to study "how" of NATO enlargement and to report in one year
December 2	Russia refuses to conclude IPP and special cooperation
December 5	President Yeltsin warns at CSCE summit in Budapest that expansion could lead to "cold peace" and the return of two blocs; suggests political membership of Russia in NATO; President Clinton affirms NATO membership open to all but NATO itself will decide on its own enlargement
December 11	Russian forces enter Chechnya
December 12	U.S. secretary of defense William Perry states NATO expansion is not accelerating, that alliance will only study "how" and "why," and that expansion is "years away"
1995	
January 5	First meeting of the NATO Senior Political Committee Reinforced to examine enlargement issues
January 16	"Intensified Dialogue" begins between NATO and partners

February 16	U.S. House of Representatives passes National Security Revitalization Act and urges as U.S. policy "that Poland, Hungary, the Czech Republic, and Slovakia . . . should be invited to become full NATO members" with option for other European countries "emerging from communist domination" also to join
April 12	Russian general Aleksandr Lebed predicts "World War III" if NATO enlarges, claims NATO has designs on territory of Russia
May 10	President Yeltsin, despite written assurance from President Clinton that Russia could join NATO, continues to oppose NATO enlargement but agrees to conclude PFP program, at Moscow summit
May 30	At Noordwijk, the Netherlands, Russia announces that it will participate in PFP and in a special "broad, enhanced dialogue" with NATO, but warns that NATO enlargement would create the need for a "corresponding correction" of Russia's attitude to PFP
September 8	President Yeltsin demands NATO become a political organization with joint European-Russian forces and warns that NATO enlargement to the borders of the Russian Federation will lead to new military blocs
September 11	Responding to NATO air attacks against Bosnian Serb targets, Russia warns NATO-Russia relationship at risk, that "luck may run out"
September 14	NATO and Ukraine launch "enhanced relationship"
September 20	NAC endorses internal enlargement study
September 28	Partners briefed in Brussels on enlargement study
October 23	At U.S.-Russian summit in Hyde Park, New York, reportedly Presidents Clinton and Yeltsin agree that Russian cooperation in Bosnia with NATO forces would delay NATO enlargement decisions
December 5	NAC decides to begin more intensive individual consultations with partners in 1996, to decide the next steps in December 1996
December 16	NATO-led Implementation Force for Yugo-

	slavia (IFOR) launched with participation of 14 partner nations
December 17	Communists win largest bloc of seats in Russian Duma

1996

January 8	Russian foreign minister Kozyrev dismissed, signaling more conservative Russian cabinet
January 12	New Russian foreign minister Yevgeni Primakov qualifies opposition to NATO enlargement by opposing eastward expansion of NATO "infrastructure" and nuclear forces
January 16	NATO inaugurates second phase of "how" and "why" consultations with partners
January 24	Polish prime minister Jozef Oleksy resigns following allegations of having spied for the Soviet Union and Russia
January 25	Council of Europe invites Russia to become member
January 25/26	Congress passes Public Law 104–107 endorsing principle of NATO enlargement but not identifying candidates or deadlines for admission or a requirement that the president expend funds to assist prospective members
January 29	Hungary first partner to request intensified bilateral consultations on admission requirements
February 9	Russian defense minister Grachev warns of possible new military alliance of CIS and Central European countries if NATO enlarges
February 22	German defense minister Rühe states that "the first negotiations on admitting new members can be expected in 1997"
February 28	Russia joins Council of Europe
February 29	Russian foreign minister Yevgeni Primakov declares: "We are not against a speedy NATO expansion, we are against expansion"
March 11	Foreign Minister Primakov alludes to possible compromise on NATO enlargement whereby NATO infrastructure would not move East
March 14	Czech Republic first partner country to submit enlargement *aide-mémoire* to NATO

March 15	Russian Duma declares dissolution of USSR illegal
March 25	President Yeltsin suggests states seeking to join NATO participate in the political committee but not the military structure
June 18	Russian Security Council Secretary Aleksandr Lebed states that "if you [NATO] have enough money and energy to expand, feel free"
June 20	Polish-UK agreement on first-ever NATO nation lease of training territory of a former Warsaw Pact member state; President Yeltsin informs President Clinton that admission of the Baltic states into NATO is "out of the question"
July 3	Boris Yeltsin reelected president of Russia by 13 percent margin
July 23	U.S. House of Representatives passes, 353 to 65, the NATO Enlargement Facilitation Act identifying the Czech Republic, Hungary, and Poland as having made most progress
July 22	Danish defense minister Hans Haekkerup predicts NATO summit in early 1997 will admit Czech Republic, Hungary, and Poland
July 25	U.S. Senate passes, 81 to 16, NATO Enlargement Facilitation Act including Slovenia
September 4	Primakov meets in Bonn to discuss concluding a NATO-Russia charter on cooperation
September 6	Secretary of State Christopher states in Stuttgart that a 1997 NATO summit "should" invite "several" partners to begin accession negotiations
October 7	Aleksandr Lebed calls for decision on NATO enlargement to be delayed for the next generation
October 22	President Clinton proposes 1999 as deadline for admitting first group of new NATO members
November 5	President Clinton reelected; Republicans retain majority in Congress
November 8	United States proposes that NACC and PFP be merged in an Atlantic Partnership Council
December 1	Thirty states parties to CFE Treaty agree to begin adaptation negotiations in 1997

December 10	NAC approves 1997 summit at which first future members will be invited to begin accession negotiations, offers Russia a "document or charter," declares no intent to deploy nuclear weapons on the territory of new members

1997

January 20	NATO secretary general Solana begins consultations in Moscow; agreed statement refers to a "successful meeting" but notes that "this will not be an easy process"
January 31	Russian prime minister Viktor Chernomyrdin states that Russia wishes to be a "full, voting member" of a NATO-Russia forum
February 3	Russian chief of staff Anatoly Chubais insists that the NATO-Russia Treaty precede the Madrid summit to avoid international and Russian domestic turmoil
February 18	U.S. secretary of state Madeleine Albright proposes joint NATO-Russia military unit, suggests that accession negotiations for new members conclude by December 1997, proposes that all partners attend Madrid summit
February 20	NATO introduces new CFE proposal in Vienna, offering lowered NATO force levels than current, pre-enlargement holdings and a special "stabilization" zone in Central Europe freezing existing levels
February 23	Second round of Solana-Primakov consultations held in Brussels; agreed statement declares "progress is emerging although differences remain"; in Helsinki, President Yeltsin predicts compromise at summit meeting with President Clinton on March 20 and 21, 1997
February 24	Clinton administration study estimates the total costs of NATO enlargement as $27 billion to $35 billion for the years 1997 to 2009
March 9	Third round of Solana-Primakov talks in Moscow
March 21	U.S.-Russia summit in Helsinki reaches elements of agreement on NATO-Russia document, including co-decision as appropriate on security issues

April 15	Fourth round of Solana-Primakov talks in Moscow
April 16	NATO proposes annual reporting on new or substantial improvement to military infrastructure
April 17	German chancellor Helmut Kohl and President Yeltsin endorse May 27 in Paris as target for concluding NATO-Russia document, citing texts as 90 percent complete
April 24	European Security Act of 1997 introduced in U.S. House calling for no special conditions for new NATO members, urging admission of the three Baltic States and Romania at earliest possible date, and reciprocity from Russia in demarcating its frontiers, stationing forces only with the consent of host states, and reducing military levels in Kaliningrad; Russian Duma compares NATO enlargement as "march to the East" that will lead to new iron curtain
April 30	NATO Individual Dialogue with 12 aspiring members concludes
May 6	Fifth round of Solana-Primakov talks in Luxembourg; parties report "some progress" and agree to intensify consultations for agreement at "earliest possible date"
May 13–14	Sixth round of Solana-Primakov talks in Moscow results in agreement on NATO-Russia document
May 27	NATO-Russia "Founding Act" signed in Paris
May 29	NATO-Ukraine Charter initialed in Sintra, Portugal
May 30	Euro-Atlantic Partnership Council created to replace NACC
July 7	President Clinton states Russia could be a future NATO member
July 8	At summit in Madrid NATO invites the Czech Republic, Hungary, and Poland to begin accession negotiations

Appendix C

Fundamental Principles of NATO Enlargement

First, potential members must be prepared to defend the Alliance and have the professional military forces to do it.

Second, NATO must continue to work by consensus—new members must respect this tradition and abide by it.

Third, military forces of new members must be capable of operating effectively with NATO forces. This means not only a common doctrine, but interoperable equipment—especially communications equipment.

Fourth, potential new members must uphold democracy and free enterprise, respect human rights inside their borders, and must respect sovereignty outside their borders.

And fifth, their military forces must be under democratic, civilian control.

—U.S. Secretary of Defense William Perry
Norfolk, Virginia, June 27, 1996

Appendix D

Signatories to the Partnership for Peace

Country	Date	Country	Date
1994			
Romania	January 26	Georgia	March 23
Lithuania	January 27	Slovenia	March 30
Poland	February 2	Azerbaijan	May 4
Estonia	February 3	Sweden	May 9
Hungary	February 8	Finland	May 9
Ukraine	February 8	Turkmenistan	May 10
Slovakia	February 9	Kazakhstan	May 27
Bulgaria	February 1	Kyrgyzstan	June 1
Latvia	February 14	Russia	June 22
Albania	February 23	Uzbekistan	July 13
Czech Republic	March 10	Armenia	October 5
Moldova	March 16		
1995			
Belarus	January 11		
Austria	February 10		
Malta	April 26 (withdrew on October 30, 1996)		
"Former Yugoslav Republic of Macedonia"	November 15		
1996			
Switzerland	December 11		
TOTAL: 27			

Appendix E

Benefits of NATO Enlargement

- Democratic reform and stability
- Stronger collective defense and ability to address new security challenges
- Improved relations among Central and East European states
- Better burdensharing and contributions to Alliance missions
- Avoidance of a destabilizing grey zone in Central and Eastern Europe
- Fostering of more stable climate for economic reform, trade, and investment
- More coherent Europe as a partner for the United States

Source: U.S. Department of State, *Report to the Congress on the Enlargement of the North Atlantic Treaty Organization: Rationale, Benefits, Costs and Implications*, February 24, 1997.

Appendix F

Focus of Pre-Accession Military Work

Near Term

- Education and training in NATO doctrine, training, and command and control
- Establishment of interoperable communications
- Identification of infrastructure needs
- Identification of requirements for integrated air defense
- Continued support of enhanced PFP initiatives

Long Term

- Participation in NATO staffs, command, and force structures
- Further improvements in interoperability
- Involvement in defense and force planning process
- Continued modernization and military reform

Source: NATO Military Committee briefing to the NATO Defense College/NAA Symposium on the Adaptation of the Alliance, Rome, April 28–30, 1997.

Notes

Introduction

1. "The Future Tasks of the Alliance (Harmel Report)," December 14, 1997, *The North Atlantic Treaty Organization: Facts and Figures* (Brussels: NATO Information Service, 1989), 403.

2. An excellent "inside" discussion of NATO's transformation in the 1990s can be found in Rob de Wijk, *NATO on the Brink of the New Millenium* (London and Washington: Brassey's, 1997).

Chapter 1

1. Eduard Shevardnadze, *The Future Belongs to Freedom* (New York: Free Press, 1991), 135–141.

2. BBC Summary of World Broadcasts (SWB), Soviet Union, September 14, 1990.

3. Speech by the president of the Czech and Slovak Federal Republic at the Parliamentary Assembly of the Council of Europe, Strasbourg, May 10, 1990.

4. *International Herald Tribune*, May 6, 1991.

5. *Defense News*, April 29, 1991.

6. *Financial Times*, March 22, 1991.

7. Quoted in Philip Zelikow and Condoleeza Rice, *Germany United and Europe Transformed: A Study in Statecraft* (Cambridge, Mass., and London: Harvard University Press, 1995), 88.

8. North Atlantic Assembly, Political Committee, *The Future of the Warsaw Pact*, Interim Report of the Subcommittee on Eastern Europe and the Soviet Union, November 1990.

9. Presentation to the USSR Diplomatic Academy, Moscow, Octo-

ber 27, 1989. Brzezinski envisaged a continued transitional presence in a united Germany of both NATO and Warsaw Pact forces. "Towards a Trans-European Commonwealth of Free Nations," *International Herald Tribune*, March 8, 1990.

10. Henry A. Kissinger, "A Plan for Europe," *Newsweek*, June 18, 1990.

11. Address by Secretary of State James Baker to the Berlin Press Club, Steigenberger Hotel, Berlin, December 12, 1989; *Current Policy*, no. 1233 (Washington, D.C.: U.S. Department of State, Bureau of Public Affairs, 1989).

12. London Declaration on a Transformed North Atlantic Alliance, issued by the Heads of State and Government participating in the meeting of the North Atlantic Council in London, July 5–6, 1990.

13. USIS Wireless File (WF), U.S. Mission NATO, July 6, 1990.

14. "The New Security Equation," Opening Address by Manfred Wörner, secretary general of the North Atlantic Treaty Organization, in Nicholas Sherwen, ed., *The Prague Conference on the Future of European Security* (NATO, 1991), 28.

15. Barbara McDougall et al., *Canada and NATO: The Forgotten Ally?* Institute for Foreign Policy Analysis, Special Report (Washington, D.C.: Brassey's, 1992), 8.

16. Willem van Eekelen, quoted in Guido Gerosa, *The North Atlantic Cooperation Council*, Interim Report of the Subcommittee on Eastern Europe and the Soviet Union Political Committee, North Atlantic Assembly, October 1991, p. 41.

17. Final Communiqué of the North Atlantic Council Ministerial meeting, Copenhagen, June 6–7, 1991.

18. "Commentary by Yuli Kvitsinski, Deputy Minister of Foreign Affairs of the USSR," in Sherwen, ed., *The Prague Conference on the Future of European Security*, 68. (Emphasis added.) The Romanian-Soviet Bilateral Treaty of Friendship and Cooperation pledged that both parties "will not participate in any way in alliances directed against each other." *Romania Libera*, April 12, 1991. Russian academics and officials would claim that other erstwhile Warsaw Pact allies refused to conclude similar clauses not because they were seeking to join NATO but because they were seeking to join the EU and its eventual security and defense identity, and that Gorbachev had been assured the larger Germany would be the last NATO real estate acquisition in the East. According to Moscow's *Sovetskaya Rossiya*, Foreign Broadcast Information Service (FBIS), Central Eurasia, June 11, 1996, "Gorbachev himself cannot say in what form or where this undertaking by the Western powers was recorded," but Russian foreign minister Primakov stated in July 1996 that French president François Mitterrand had made this pledge in 1990–1991 along with "many Western leaders." *Trud*, June 25, 1996, BBC SWB Former

USSR, June 28, 1996. In 1997 former Soviet foreign minister Eduard Shevardnadze, the current Georgian president, stated that no Western guarantees had ever been extended regarding the nonenlargement of NATO following German unification. Reuters, February 14, 1997. However, then U.S. ambassador to Moscow Jack Matlock maintains that "when Gorbachev and others say that it is their understanding NATO expansion would not happen, there is a basis for it." And Secretary of State Baker is reported to have stated on February 8, 1990, in a meeting with Gorbachev that "there would be no extension of NATO's current jurisdiction eastward." Michael R. Gordon, "Did Gorbachev Get U.S. NATO Pledge?" *International Herald Tribune*, May 26, 1997.

19. "The Situation in the Soviet Union," statement issued by the North Atlantic Council in Ministerial Session, Brussels, May 21, 1991.

20. McDougall, *Canada and NATO*, 9.

21. USIS WF, October 4, 1991.

22. USIS WF, December 23, 1991.

23. Quoted in Gerosa, *The North Atlantic Cooperation Council*, Interim Report of the Subcommittee on Eastern Europe and the Former Soviet Union, p. 5.

24. Declaration, Extraordinary Meeting of the WEU Council of Ministers with the States of Central Europe, Bonn, June 19, 1992.

25. "The Document of Russia's Foreign Policy," FBIS, Soviet Union, December 2, 1992.

26. USIS WF, May 5, 1992.

27. Gerosa, *The North Atlantic Cooperation Council*, Interim Report of the Subcommittee on Eastern Europe and the Former Soviet Union, 16.

28. Ambassador Reginald Bartholomew, "The Atlantic Community after Communism," 8th Annual Conference for Director of Strategic Studies Institutes, Knokke-Heist, Belgium, September 23, 1992.

29. Ibid. Polish Lt. Colonel Grzegorz Wisniewski proposed creating a Visegrad NATO-standard air traffic control and air defense system, leasing AWACS, and actively integrating into NATO—which he viewed as an urgent task should Russia seek again to dominate the region. "Polish Security Doctrine," manuscript, January 1992. Later that year Poland offered bases for peacekeeping training (*The European*, November 27, 1992), and at the November 1992 NAA Annual Session in Brugge, Belgium, Hungary (feeling exposed because of the war in the neighboring former Yugoslavia) offered to host NATO rapid reaction forces.

30. USIS WF, December 22, 1992.

31. Andrzej Karkoszka, "Security Policy and the Armed Forces of Poland," unpublished manuscript, March 1993.

Chapter 2

1. *Defense News*, July 13, 1992.

2. USIS WF, May 5, 1992.

3. Philip Zelikow, "NATO Expansion Wasn't Ruled Out," *International Herald Tribune*, July 6, 1995.

4. Jeffrey Simon, "Does Eastern Europe Belong in NATO?" *Orbis* (Winter 1993): 19–35. Likewise, Rand Corporation analysts, in another pioneering article, suggested "association agreements" spelling out NATO membership criteria to provide a clear perspective. Ronald D. Asmus, Richard L. Kugler, and F. Stephen Larrabee, "Building a New NATO," *Foreign Affairs* (September/October 1993), 36.

5. Gerosa, *The North Atlantic Cooperation Council*, Interim Report of the Subcommittee on Eastern Europe and the Former Soviet Union, 11.

6. FBIS Western Europe, November 1, 1993.

7. FBIS Western Europe, October 5, 1993.

8. Richard Lugar, "NATO's Near Abroad," address to the U.S. Atlantic Council, Washington, D.C., December 9, 1993.

9. Jamie Shea, "NATO's Eastern Dimension," presentation at the 9th Annual Strategic Studies Conference, Knokke-Heist, Belgium, September 23–26, 1993.

10. Ibid.

11. Ibid.

12. The Rt. Hon. Lord Carrington, "The Liberation of Europe: Safeguarding the Legacy," *RUSI Journal* (August 1994): 12.

13. Hermann Freiherr von Richthofen, "Transatlantic Security Relations and the Development of a European Security and Defense Identity," presentation at the 9th Annual Strategic Studies Conference, Knokke-Heist, Belgium, September 23–26, 1993.

14. Longin Pastusiak, *A View from Poland on the New European Security Order*, Special Report of the Working Group on the New European Security Order, Political Committee, North Atlantic Assembly, October 1993.

15. Prague *Mlada Fronta Dnes*, December 2, 1993, in FBIS Eastern Europe, December 2, 1993.

16. Speech by the secretary general of NATO to the International Institute for Strategic Studies, Brussels, September 10, 1993. The speech was given under the assumption that the Polish-Russian communiqué did indeed signal a green light from Moscow on NATO enlargement.

Chapter 3

1. Wojciech Lamentowicz, "Future Political Structure of Europe Seen from an Eastern Point of View," presentation at the NATO Defense College, Rome, October 24, 1990.

2. General James McCarthy, "Opportunities for Strengthening Security in Central and Eastern Europe," presentation at the 8th Annual Conference for Directors of Strategic Studies Institutes, Knokke-Heist, Belgium, September 24, 1992.

3. Ibid.

4. Michael Dobbs, "Turmoil Marks Debate over NATO's Makeup and Mandate," *International Herald Tribune*, July 6, 1995.

5. Richard Lugar, "NATO's 'Near Abroad': New Membership, New Missions," address to the U.S. Atlantic Council, Washington, D.C., December 9, 1993.

6. USIS WF, August 16–17, 1993.

7. Presentation by the Honorable Volker Rühe, minister of defense, Federal Republic of Germany, to the North Atlantic Assembly Defense and Security Committee, Berlin, May 21, 1993.

8. Reuters, October 21, 1993.

9. Sir David Gillmore, "Representing Britain Overseas: Post-Cold War Challenges," *RUSI Journal* (December 1993): 15.

10. FBIS Western Europe, October 25, 1993.

11. FBIS Western Europe, October 20, 1993.

12. *International Herald Tribune*, October 8, 1993.

13. *The Times*, October 8, 1993.

14. Moscow *Perspektivy Rashireniya* (1993), in FBIS, Former Soviet Union, December 8, 1993.

15. USIS WF, October 22, 1993.

16. Ambassador Giovanni Jannuzzi, "Partnership for Peace: Where Do We Stand—The NATO View," presentation at the 10th Annual Strategic Studies Conference, Knokke-Heist, Belgium, September 15–18, 1994.

17. USIS WF, December 6, 1993.

18. Ibid.

19. *Atlantic News*, November 5, 1993.

20. *Frankfurter Allgemeine*, September 30, 1993.

Chapter 4

1. Statement of Admiral William D. Smith, United States Representative to the NATO Military Committee, before the Subcommittee on Coalition Defense and Reinforcing Forces, Senate Armed Services Committee, June 18, 1993.

2. "The NATO Summit and the Future of European Security," statement of Hon. Stephen A. Oxman, assistant secretary of state for European and Canadian Affairs, before the Subcommittee on Coalition Defense of the Senate Armed Services Committee and the Subcommittee on Europe of the Senate Foreign Relations Committee, February 1, 1994.

3. FBIS Western Europe, December 9, 1993.

4. For President Clinton's remark, see USIS WF, July 11, 1994; for Secretary Christopher's, see USIS WF, June 1, 1994.

5. FBIS Western Europe, December 17, 1993.

6. *NATO's Partnership for Peace: Views from Warsaw and Moscow*, North Atlantic Assembly Staff Trip Report, Document AL 66 PC/EE (94) 1, May 1994. Emphasis added.

7. BBC SWB Former USSR, January 14, 1994.

8. "An Initial Look at Reaction to Partnership for Peace," SHAPE Central and East European Defense Studies, January 14, 1994.

9. BBC SWB Eastern Europe, January 13, 1994.

10. Ibid.

11. FBIS Central Eurasia, January 13, 1994.

12. Robert Zoellick, "Set Criteria for NATO Membership Soon," *International Herald Tribune*, January 7, 1994.

13. Henry Kissinger, "At Sea in a New World," *Newsweek*, June 6, 1994.

14. Henry Kissinger, "Be Realistic about Russia," *Washington Post*, January 25, 1994.

15. Zbigniew Brzezinski, "The Way Forward for an Inspired NATO," *International Herald Tribune*, December 2, 1993.

16. Senator Richard Lugar, "European Security Revisited," presentation at the Overseas Writers Club, Washington, D.C., June 28, 1994. Emphasis added.

17. Peter Corterier, Secretary General of the North Atlantic Assembly, "Meeting NATO's New Challenges," graduation speech at the NATO Defense College, February 11, 1994.

Chapter 5

1. Andranik Migranyan, "Unequal Partnership," *New York Times*, June 23, 1994.

2. Sergei Karaganov, "We Must Be the First At NATO's Doors," *Izvestiya*, February 24, 1994, in FBIS Central Eurasia, March 1, 1994.

3. Statement Made on January 12, 1994, at a Briefing in the Press Center of the Foreign Ministry of Russia Concerning the Results of the NATO Summit in Brussels.

4. BBC SWB Former USSR, March 4, 1994.

5. G. Yevgenyev, I. Kirin, *USA-Western Europe: Flare-Up Of Old Discord* (Moscow: Novosti Press Agency Publishing House, 1975), 12.

6. *Atlantic News*, March 12, 1994, and March 2, 1994.

7. FBIS Central Eurasia, April 14, 1994.

8. BBC SWB Former USSR, March 22, 1994.

9. BBC SWB Former USSR, April 21, 1994.
10. FBIS Central Eurasia, May 23, 1994.
11. USIS WF, June 24, 1994.

Chapter 6

1. FBIS Eastern Europe, June 29, 1994.
2. Ibid. Emphasis added.
3. "Central European Security, 1994: Partnership for Peace," *Strategic Forum* (1:1994).
4. Author's notes, June 21, 1994.
5. Zbigniew Brzezinski, "Improved U.S. Policy for Russia and Central Europe," *International Herald Tribune*, June 29, 1994.
6. USIS WF, July 7, 1994.
7. USIS WF, July 8, 1994. Emphasis added.
8. Tomasz Wroblewski, "Poland-United States, with Russia in the Background," Warsaw *Zycie Warszawy*, July 7, 1994, in FBIS Eastern Europe, July 7, 1994.
9. FBIS Eastern Europe, July 12, 1994.
10. FBIS Central Eurasia, July 6, 1994.
11. Frederick Kempe, "The Answer to Europe's German Question," *Wall Street Journal Europe*, September 16–17, 1994.
12. *International Herald Tribune*, September 10, 1994.
13. David Williams and R. Jeffrey Smith, "NATO Inching toward East," *International Herald Tribune*, November 7, 1994. UK defense minister Malcolm Rifkind reportedly stated that Russia in NATO was "not a serious proposition," whereas Turkey, concerned that Russian consent to NATO enlargement in Central Europe might entail concessions to Russian influence elsewhere, supports NATO membership for Albania, Bulgaria, and Romania. Bruce Clark, "NATO Rallies to Call of 'Expand or Die,'" *Financial Times*, September 29, 1994.
14. Cited in Wendy Ross, "Surplus U.S. Military Equipment Can Be Sold to Eastern Europe," U.S. Information Service, October 14, 1994.
15. Williams and Smith, "NATO Inching toward East."
16. North Atlantic Assembly, *General Report of the Political Committee: Annex*, November 1994.
17. For elaboration *see* Zbigniew Brzezinski, "A Plan for Europe," *Foreign Affairs* (January/February 1995): "It is not carping criticism to point out that, so far, the Clinton administration has projected neither a strategic vision nor a clear sense of direction on a matter of such salience to Europe's future as enlarging NATO," 27. NATO Defense Procurement Policy officer Brent Fischmann stated on June 2, 1994, that participation in the PFP "does not require . . . the wholesale modernization of

the equipment of the Czech Armed Forces. We know from the Gulf War that cooperation is possible between our forces when critical interoperability issues are addressed," which he identified as including tactical communications, the ability to share tactical reconnaissance and other data, cross-servicing of aircraft, IFF, instrument landing systems, self protection of transport aircraft and armored vehicles, mine detection and destruction." "Partnership for Peace: NATO's Approach to Expanding Defense-Related Cooperation," presentation at the conference "The Czech Republic on Her Way to the Western Alliance," Brno, Czech Republic, June 2, 1994. Indeed, Soviet aircraft had actually been designed to be serviced at NATO airfields on the assumption that a conflict would result in a rapid occupation of Western Europe!

18. North Atlantic Assembly, Presentation to the Political Committee, Washington, D.C., November 15, 1994.

19. North Atlantic Assembly, Plenary Resolution, Washington, D.C., November 18, 1994.

20. *The Independent*, November 22, 1994.

21. *The Independent*, October 31, 1994.

22. Financial Times, December 22, 1994.

23. BBC SWB Former USSR, December 6, 1994.

24. BBC SWB Former USSR, December 22, 1994.

25. USIS WF, January 13, 1995.

Chapter 7

1. *North Atlantic Treaty*, Report of the Committee on Foreign Relations, U.S. Senate, Executive Report no. 8, 81st Congress, 1st session, June 6, 1949, p. 28.

2. "Text of the Report of the Committee of Three on Non-Military Cooperation in NATO," approved by the North Atlantic Council on December 13, 1956, *The North Atlantic Treaty Organization: Facts and Figures* (Brussels: NATO Information Service, 1989), 385, 387.

3. "A future adversary would never believe that the United States would risk its own survival to extend the nuclear umbrella in defense of nations where it has little economic, political or security interests. . . . The Balkans war has set the precedent with the United States' refusal to become involved and our allies' rejection of military force to defend interests on their own frontiers against a comparatively weak opponent." Representative Robert G. Torricelli (D-N.J.), "Perspective on NATO Expansion: A Promise Best Not Kept," *Los Angeles Times*, February 9, 1995.

4. Robert Pszczel, "Polish Perceptions of the Partnership for Peace Initiative," *International Defense Review* (Defense '95 issue), 19–21.

5. Longin Pastusiak, *Poland on Her Way into NATO*, Draft Special Report of the Subcommittee on Transatlantic and European Relations, Political Committee, North Atlantic Assembly, April 23, 1997, p. 1.

6. Alexei Pushkov, "Reacting to NATO Expansion, Russia Should Take Its Time," *Moscow News*, March 24–30, 1995. These conditions were said to include extending enlargement in space and time, nondeployment of nuclear weapons and foreign troops on a permanent basis in peacetime, a ban on military exercises involving foreign troops on territories contiguous to Russia "without Moscow's consent," a ban on the development of offensive weapons in these areas, notification of troop movements above agreed limits, and guarantees of the *status quo* in Kaliningrad oblast. Yuri Davidov, "Moscow Urged to Accept NATO Expansion at Price," FBIS Central Eurasia, March 24, 1995. For example, Russian naval commander Feliks Gromov observed with reference to Poland and the Baltics joining NATO: "It's one thing just to join without any troops being stationed. It's another thing for NATO troops to be stationed." FBIS Central Eurasia, April 26, 1995.

7. *Sunday Telegraph*, May 14, 1995.

8. *Defense News*, June 4, 1995.

9. Eduard Rizhkin, "Less than Half of Moscow Interviewees Are against NATO Expansion," *Segodnya*, June 9, 1995; 25.3 percent were for Eastern Europe in NATO, 44.3 percent against, and 24 percent were neutral. Former prime minister Yegor Gaidar stated: "Russia has so many problems and worries that this one [NATO enlargement] is really not the most important." FBIS Eastern Europe, March 15, 1995.

10. FBIS Central Eurasia, April 13, 1995.

11. Ibid.

12. FBIS Western Europe, May 16, 1995.

13. *Wall Street Journal Europe*, May 15, 1995.

14. BBC SWB Former USSR, May 12, 1995.

15. *International Herald Tribune*, May 31, 1995.

16. Interfax, June 1, 1995.

17. Senator Sam Nunn, "The Future of NATO in an Uncertain World," speech to the SACLANT Seminar 95, June 22, 1995, Norfolk, Virginia.

18. Jan Petersen, *Towards a Security Strategy for NATO and Europe*, General Report of the Political Committee, North Atlantic Assembly, October 1995, pp. 16–17. Petersen was quoting Mary N. Hampton, "U.S. Foreign Policy, West Germany and the Wilsonian Impulse," *Security Studies* (Spring 1995): 610–611.

19. House of Commons Defence Committee, *The Future Of NATO: The 1994 Summit and Its Consequences*, Tenth Report, July 1995.

20. *Russia and NATO: Theses of the Council on Foreign and Defense*

Policy, published in *Nezavisimaya Gazeta,* June 21, 1995, in FBIS Central Eurasia, June 23, 1995.

21. Sergei Rogov, "NATO's Enlargment: The Unresolved Issues," presentation to the 11th Annual Strategic Studies Conference *NATO and the New Security Architecture,* sponsored by U.S. Mission NATO and the Institute for National Security Studies of the U.S. National Defense University, Knokke-Heist, Belgium, September 7–10, 1995.

22. Willem van Eekelen, *The Security Agenda for 1996: Background and Proposals* (Brussels: Center for European Policy Studies, 1995), 34.

23. Peter W. Rodman, "Is the Iron Curtain Gone or Not?" *International Herald Tribune,* January 18, 1996.

Chapter 8

1. *Atlantic News,* December 8, 1995.
2. USIS *WF,* October 12, 1995.
3. FBIS Central Eurasia, December 9, 1995.
4. BBC SWB Former USSR, January 29, 1996.
5. *Washington Times,* January 24, 1996.
6. *International Herald Tribune,* February 5, 1996.
7. USIS *WF,* January 31, 1996.
8. BBC SWB Former USSR, February 21, 1996.
9. BBC SWB Former USSR, February 27, 1996.
10. Letter to Jan Nowak, February 13, 1996, cited with permission.
11. *Moscow News,* January 19–25, 1996.
12. Reuters, March 12, 1996.
13. March 1994 statement from BBC SWB Former USSR, March 26, 1996; Gorbachev proposal cited in Zelikow and Rice, *Germany United and Europe Transformed,* 270.
14. Robert B. Zoellick, Statement on NATO enlargement before the U.S. Senate Foreign Relations Committee, April 27, 1995.
15. *Le Monde,* June 3, 1996.
16. Ambassador Gebhardt von Moltke, "NATO and European Security," *Central European Issues* (Summer 1996).
17. Nicholas Williams, "The Future of Partnership for Peace," Arbeitspapier, Konrad Adenauer Stiftung, April 1996.
18. Robert Pszczel, "The Enlargement of NATO and the Future of Peacekeeping," *Verification 1996,* J. B. Poole and R. Guthrie, eds. (Boulder: Westview, 1996), 301.
19. FBIS Eastern Europe, June 24, 1996.
20. Secretary of State Warren Christopher, "A New Atlantic Community for the 21st Century," USIS WF, September 6, 1996.

21. BBC SWB Former USSR, September 4, 1996.

22. *International Herald Tribune*, October 11, 1996.

23. *Guardian*, October 8, 1996.

24. Cited in Michael Mihalka, "The Emerging European Security Order," *Transition*, December 15, 1995, p. 17.

25. FBIS Central Eurasia, July 29, 1996.

26. FBIS Central Eurasia, August 2, 1996.

27. BBC SWB Former USSR, October 11, 1996.

28. Colonel General Igor Rodionov, "Russia and NATO: After Bergen," *Moscow News*, October 3–9, 1996.

29. BBC SWB Former USSR, June 20, 1996.

30. U.S. Congressional Budget Office, *The Costs of Expanding the NATO Alliance*, March 1996.

31. Ronald D. Asmus, Richard L. Kugler, and F. Stephen Larrabee, "What Will NATO Enlargement Cost?" *Survival* (Autumn 1996): 5–26. The estimates by these Rand Corporation analysts suggested a lower figure of $10–110 billion for enlarging to the same four countries over a 10 to 15 year period, and assumed a non-worst case scenario.

32. USIS WF, September 5, 1996.

33. *Congressional Record—House*, July 23, 1996, p. H8118.

34. USIS WF, October 4, 1991.

35. USIS WF, October 22, 1996.

36. Sir John Goulden KCMG, "NATO Approaching Two Summits," presentation at the Royal United Services Institute for Defence, October 3, 1996.

Chapter 9

1. Final Communiqué, Ministerial Meeting of the North Atlantic Council, Brussels, December 10, 1996, in *NATO Review* (January 1997): 31–35.

2. USIS WF, July 18, 1996.

3. "Political Implications of NATO Enlargement," Presentation by Gebhardt von Moltke, assistant NATO secretary general for political affairs, to the NATO Defense College/NAA Symposium on the Adaptation of the Alliance, Rome, April 28, 1997.

4. Bruce George, *Complementary Pillars of European Security: The Organization for Security and Cooperation in Europe and the Atlantic Partnership Council*, Draft Interim Report of the Subcommittee on Transatlantic and European Organizations, North Atlantic Assembly, March 20, 1997, p. 7.

5. Presentation by Major General Joël Marchand, assistant director for cooperation and regional security, NATO International Military Staff, to the NATO Defense College/NAA Symposium on the Adaptation of the Alliance, Rome, April 29, 1997.

6. "Military Dimensions of NATO Enlargement," address by General Klaus Naumann, chairman of the NATO Military Committee, to the North Atlantic Assembly in Brussels, February 16, 1997.

7. Cited in North Atlantic Assembly, *Baltic Watch: Seminar on Baltic Security Perspectives*, Riga, Latvia, December 8–10, 1996, pp. 1, 2.

8. USIS WF, June 25, 1996.

9. USIS WF, November 26, 1996.

10. Jan Petersen, *Defining Moments: Alliance Developments 1996*, General Report of the Political Committee, North Atlantic Assembly, November 1996, p. 16.

11. NAA, *Baltic Watch: Seminar on Baltic Security Perspectives*.

12. *Baltic Times*, April 3–9, 1997.

13. Robert Blackwill, Arnold Horelick, and Sam Nunn, *Stopping the Decline in U.S.-Russian Relations*, Rand Paper P-7986 (1996), 3.

14. BBC SWB Former USSR, April 2, 1997.

15. USIS WF, February 18, 1997.

16. BBC SWB Former USSR, March 17, 1997. Emphasis added.

17. BBC SWB Former USSR, March 19, 1997.

18. "Joint U.S.-Russian Statement on European Security," USIS WF, March 21, 1997.

19. BBC SWB Former USSR, March 24, 1997.

20. *International Herald Tribune*, May 9, 1997.

21. Henry Kissinger, "Helsinki Fiasco," *Washington Post*, March 30, 1997.

22. Dissenting View, *Task Force Report: Russia, Its Neighbors, and an Enlarging NATO*, sponsored by the Council on Foreign Relations, 1997, p. 28.

23. Maarten van Traa, *Towards the NATO Madrid Summit*, Draft General Report of the Political Committee, North Atlantic Assembly, May 1, 1997, p. 6.

24. NATO Office of Information and Press, "The Alliance's Strategic Concept," Appendix IX in *NATO Handbook* (Brussels: NATO Office of Information and Press, 1995), 239.

25. BBC SWB Former USSR, February 5, 1996.

26. BBC SWB Former USSR, October 25, 1996.

27. USIS WF, April 23, 1997.

28. Ibid.

29. Von Moltke, "Political Implications of NATO Enlargement."

30. Congressman Gerald B. Solomon, "Prizes and Pitfalls of NATO Enlargement," *Orbis* (Spring 1997): 220.

31. *Atlantic News*, May 23, 1997. For a brief history see the interview with Secretary General Solana by Xavier Vidal-Foch, "The Pact with Russia Puts an End to a Divided Europe," *El Pais*, May 18, 1997.

32. BBC SWB Former USSR, May 19, 1997.

33. Reuters, May 7, 1997.

34. Poll conducted by the Democratic Initiatives Center, UNIAN news agency, May 6, 1997, cited by Reuters, May 7, 1997.

35. *International Herald Tribune*, May 23, 1997.

36. "Military Dimensions of NATO Enlargement," address by General Klaus Naumann, February 16, 1997.

37. USIS WF, April 23, 1997.

38. Cited in Alan K. Henrikson, "The Creation of the North Atlantic Alliance," *American Defense Policy*, fifth edition, John F. Reichart and Steven R. Sturm, eds. (Baltimore: Johns Hopkins Press, 1982), 300.

39. George F. Kennan, "NATO Expansion Would Be a Fateful Blunder," editorial, *International Herald Tribune*, February 6, 1997.

40. Michael Mandelbaum, *The Dawn of Peace in Europe* (New York: Twentieth Century Fund Press, 1996), 55.

41. USIS WF, January 9, 1997.

42. USIS WF, March 31, 1997.

43. USIS WF, April 23, 1997.

44. "Recommendation 601 on Defence and Security in an Enlarged Europe," in WEU Assembly, *Proceedings*, Assembly of the Western European Union, 42d session, December 1996, p. 25.

45. *Atlantic News*, March 26, 1997.

46. Annette Just and Porter J. Goss, *Not Whether, Not When, But Now*, Draft Interim Report of the Subcommittee on NATO Enlargement and the New Democracies, North Atlantic Assembly, May 9, 1997, pp. 5-7.

47. *International Herald Tribune*, January 21, 1997. Another poll showed 37 percent opposing NATO enlargement to Central Europe, 14 percent in favor, but 49 percent with no opinion, whereas another poll showed only 18 percent believing the government's opposition as intended to protect Russian security and 38 percent believing this stance to be a tactic to get more foreign aid, divert attention from domestic problems, or influence domestic politics. *International Herald Tribune*, May 29, 1997.

48. Center of International Sociological Research, "Public Opinion in Russia on Lithuania Joining NATO," March 1997.

49. *Americans on Expanding NATO: A Study of U.S. Public Attitudes* (College Park, Md.: University of Maryland, October 1, 1996).

50. *Krasnaya Zvezda*, February 25, 1997.

51. Heinrich Vogel, "Opening NATO: A Cooperative Solution for an Ill-Defined Problem?" *Aussenpolitik* 1 (1997): 28.

52. Address by Javier Solana, secretary general of NATO, at the University of Warsaw, April 18, 1996.

53. Letter to the director of the Political Committee, North Atlantic Assembly, May 12, 1997. Quotation from *Washington Post*, March 13, 1997.

Chapter 10

1. Remarks by President Boris Yeltsin, NATO-Russia Founding Act Signing Ceremony, May 27, 1997.

2. Tatiana Parkhalina, "Of Myths and Illusions: Russian Perceptions of NATO Enlargement," *NATO Review* (May–June 1997): 12. Emphasis added. Nevertheless, the Russian ambassador to the United Kingdom Anatoli Adamshin asserted that "at least 90 percent of Russians are against enlargement." *Daily Telegraph*, May 20, 1997. For instance, a March 1997 poll showed 70 percent of Russian public opinion as favoring Lithuanian neutrality, but 59 percent opposed Lithuania's serving as a "buffer zone," and 44 percent agreed that NATO membership would increase Lithuanian security. Center of International Sociological Research, "Public Opinion in Russia on Lithuania Joining NATO," March 1997.

3. *Daily Telegraph*, May 18, 1997.

4. Javier Solana, "NATO's Quantum Leap," *Wall Street Journal Europe*, May 28, 1997. According to Sergei Rogov, the director of the USA and Canada Institute of the Russian Academy of Sciences, the act "is a geopolitical arrangement that allows Russia to avoid the status of a defeated nation." *Washington Times*, June 4, 1997.

5. Zbigniew Brzezinski, "The Germ of a More Secure Europe," *Financial Times*, May 27, 1997.

6. "Military Dimensions of NATO Enlargement," address by General Klaus Naumann, February 16, 1997.

7. *Atlantic News*, May 31, 1997; *Defense News*, June 9–15, 1997.

8. Klaus Kinkel, "The Resolute NATO and EU Goal Is Security for All of Europe," *International Herald Tribune*, May 30, 1997. However, the foreign minister rejected a "special status" or bilateral security guarantees for the Baltic states.

9. Reuters, June 18, 1997.

10. USIS WF, May 29, 1997.

11. Ibid.

12. Floyd D. Spence and Ronald V. Dellums, "Is a Bigger NATO Also Better?" *Washington Times*, May 29, 1997.

13. Separately, Senator Helms challenged much of NATO's transformation in the 1990s, encapsulated by Senator Richard Lugar's highly

popularized June 1993 call "out of area or out of business," arguing its role was not to replace the UN as "the world's peacekeeper" or "build democracy and pan-European harmony or promote better relations with Russia." Rather, Chairman Helms argued that a "central strategic rationale" for enlargement must be to hedge against the possible return of a nationalist or imperialist Russia. Jesse Helms, "New Members, Not New Missions," *Wall Street Journal*, July 9, 1997. In contrast, F. Stephen Larrabee argues that enhancing NATO's ability to conduct "out-of-area" crisis management missions as in IFOR/SFOR *viz.* collective defense would open up new possibilities for cooperation with Russia and reduce Russian suspicions of NATO—although no doubt extremist politicians would point to NATO preparations for intervention in the former Soviet Union. F. Stephen Larrabee, *NATO Enlargement and the Post-Madrid Agenda*, RAND Paper P-7999 (1997).

14. USIS WF, June 2, 1997.
15. USIS WF, June 12, 1997.
16. BBC SWB Former USSR, July 10, 1997.

Selected Bibliography

Americans on Expanding NATO: A Study of U.S. Public Attitudes. College Park, Md.: University of Maryland, October 1996.

Ananicz, Andrzej, Premyslaw Gruzinski, Andrzej Olechowski, Janusz Onyszkiewicz, Krzysztof Skubiszewski, and Henryk Szlajger. "Poland-NATO Report." *European Security* (Spring 1996): 141–166.

Asmus, Ronald D., Richard L. Kugler, and F. Stephen Larrabee. "Building a New NATO." *Foreign Affairs* (September/October 1993): 28–40.

———. "What Will NATO Enlargement Cost?" *Survival* (Autumn 1996): 5–26.

Biden, Joseph. *Meeting the Challenges of a Post-Cold War World: NATO Enlargement and U.S.-Russia Relations.* Report to the Committee on Foreign Relations, U.S. Senate, 105th Congress, 1st session, May 1997. Washington, D.C.: U.S. Government Printing Office, 1997.

Blackwill, Robert, Arnold Horelick, and Sam Nunn. *Stopping the Decline in U.S.-Russian Relations.* Rand Paper P-7986. Santa Monica, Calif.: RAND, 1996.

Borawski, John. "If Not NATO Enlargement: What Does Russia Want?" *European Security* (Autumn 1995): 381–395.

———. *The NATO-Russia Founding Act.* Briefing Paper no. 12. Brussels: International Security Information Service, July 1997.

———. "Partnership for Peace and Beyond." *International Affairs* (April 1995): 233–246.

Borawski, John, and Masha Khmelevskaja. "Why NATO Enlargement? Why NATO?" *Central European Issues* 2, no. 1 (1996): 61–71.

Brzezinski, Zbigniew. "A Plan for Europe." *Foreign Affairs* (January/February 1995): 26–42.

Butcher, Martin, Tasos Kokkinides, and Daniel Plesch. *Study on NATO Enlargement: Destabilizing Europe.* Joint Report of the British American Information Council and the Center for European Security and Disarmament, 1995.

Congressional Budget Office (CBO). *The Costs of Expanding the NATO Alliance*. Washington, D.C.: CBO, March 1996.

Cooperation and Partnership for Peace: A Contribution to Euro-Atlantic Security into the 21st Century. Whitehall Paper. London: Royal United Services Institute for Defence Studies, 1996.

Council on Foreign Relations. *Russia, Its Neighbors, and an Enlarging NATO*. Report of an Independent Task Force. New York: Council on Foreign Relations, May 5, 1997.

Gerosa, Guido. "The North Atlantic Cooperation Council." *European Security* (Autumn 1992): 273–294.

Goldgeier, James M. "NATO Expansion: The Anatomy of a Decision." *Washington Quarterly* 21, no. 1 (Winter 1998): 85–102.

Grimmett, Richard F., Paul E. Gallis, and Larry Nowels. *NATO: Alliance Expansion, Partnership for Peace, and U.S. Security Assistance*. Congressional Research Service (CRS), Report for Congress no. 97–531F (Washington, D.C.: CRS, Library of Congress, May 9, 1997).

Haglund, David G., ed. *Will NATO Go East? The Debate over Enlarging the Atlantic Alliance*. Kingston, Ontario: The Centre for International Relations, Queen's University, 1996.

Henrikson, Alan K. "The Creation of the North Atlantic Alliance." In *American Defense Policy*, edited by John F. Reichart and Steven R. Sturm, 296–320. Fifth edition. Baltimore: Johns Hopkins Press, 1982.

Holbrooke, Richard. "America, A European Power." *Foreign Affairs* (March/April 1995): 38–51.

Johnsen, William T., and Thomas-Durell Young. *Partnership for Peace: Discerning Fact from Fiction*. Carlisle Barracks, Pa.: Strategic Studies Institute, U.S. Army War College, 1994.

Just, Annette, and Porter J. Goss. *Ratification Of NATO Enlargement*. Interim Report of the Subcommittee on NATO Enlargement and the New Democracies, Political Committee, North Atlantic Assembly, August 1997.

———. *Not Whether, Not When, But Now*. Draft Interim Report of the Subcommittee on NATO Enlargement and the New Democracies, Political Committee, North Atlantic Assembly, May 1997.

Kugler, Richard. *Costs of NATO Enlargement: Moderate and Affordable*. Strategic Forum no. 128. Washington, D.C.: Institute for National Strategic Studies, National Defense University, October 1997.

Lamentowicz, Wojciech. "Future Political Structure of Europe Seen from an Eastern Point of View." Presentation at the NATO Defense College, Rome, October 24, 1990.

Larrabee, F. Stephen. *NATO Enlargement and the Post-Madrid Agenda*, RAND Paper P-7999. Santa Monica, Calif.: RAND, 1997.

Lugar, Richard. "NATO's 'Near Abroad': New Membership, New Missions." Address to the U.S. Atlantic Council. Washington, D.C., December 9, 1993.

_____. "European Security Revisited." Presentation at the Overseas Writers Club, Washington, D.C., June 28, 1994.

Mandelbaum, Michael. *The Dawn of Peace in Europe.* New York: Twentieth Century Fund Press, 1996.

McCarthy, James. "Opportunities for Strengthening Security in Central and Eastern Europe." Presentation at the 8th Annual Conference for Directors of Strategic Studies Institutes. Knokke-Heist, Belgium, September 24, 1992.

Medvedev, Sergei. *NATO Enlargement: Russian Perspectives.* Paper SWP-AP 2965. Ebenhausen, Germany: Stiftung Wissenschaft und Politik, July 1996.

NATO Enlargement: Cost Estimates Developed to Date Are Notional. Washington, D.C.: U.S. General Accounting Office, August 18, 1997.

NATO Enlargement Watch. North Atlantic Assembly newsletter (quarterly).

NATO Review. Summit edition. July–August 1997.

North Atlantic Treaty. Report of the Committee on Foreign Relations, U.S. Senate, Executive Report no. 8, 81st Congress, 1st session, June 6, 1949. Washington, D.C.: U.S. Government Printing Office, 1949.

North Atlantic Treaty Organization. *Study on NATO Enlargement.* September 1995.

Nunn, Sam. "The Future of NATO in an Uncertain World." Speech to the SACLANT Seminar 95, Norfolk, Virginia, June 22, 1995.

Parkhalina, Tatiana. "Of Myths and Illusions: Russian Perceptions of NATO Enlargement." *NATO Review* (May–June 1997): 11–15.

Pastusiak, Longin. *Poland on Her Way to NATO.* Special Report of the Subcommittee on Transatlantic and European Relations, Political Committee, North Atlantic Assembly, September 1997.

Petersen, Jan. *Towards a Security Strategy for NATO and Europe.* General Report, Political Committee, North Atlantic Assembly, October 1995.

Pszczel, Robert. "The Enlargement of NATO and the Future of Peacekeeping." In *Verification 1996,* edited by J. B. Poole and R. Guthrie, 292–304. Boulder: Westview, 1996.

Report to the Congress on the Enlargement of the North Atlantic Treaty Organization: Rationale, Benefits, Costs and Implications. Washington, D.C.: Bureau of European and Canadian Affairs, U.S. Department of State, February 24, 1997.

Rogov, Sergei. "NATO's Enlargement: The Unresolved Issues." Presentation to the 11th Annual Strategic Studies Conference, *NATO and* the *New Security Architecture*, sponsored by U.S. Mission NATO and the Institute for National Security Studies of the U.S. National Defense University. Knokke-Heist, Belgium, September 7–10, 1995.

Rosner, Jeremy D. "NATO Enlargement's American Hurdle." *Foreign Affairs* (July–August 1997): 9–16.

_____. "The American Public, Congress and NATO Enlargement." *NATO Review* (January 1997): 12–14.

Roth, William V. "A Fresh Act of Creation: The Parliamentary Dimension of NATO Enlargement." *NATO Review* (March 1997): 11–13.

Rudolf, Peter. "The Future of the United States as a European Power: The Case of NATO Enlargement." *European Security* (Summer 1996): 175–195.

"Russia and NATO: Theses of the Council on Foreign and Defense Policy." Moscow *Nezavisimaya Gazeta*, June 21, 1995. In *Foreign Broadcast Information Service*, Former Soviet Union, June 23, 1995, pp. 10–18.

Russian Federation Foreign Intelligence Service. "Prospects of NATO Expansion and Russia's Interests." Moscow *Perspektivy Rashireniya* (1993). In *Foreign Broadcast Information Service*, Former Soviet Union, December 8, 1993, pp. 61–71.

Sanz, Timothy. "NATO's Partnership for Peace Program: Published Literature." *European Security* (Winter 1995): 676–696.

Shea, Jamie. "NATO's Eastern Dimension." Presentation at the 9th Annual Strategic Studies Conference. Knokke-Heist, Belgium, September 23–26, 1993.

Simon, Jeffrey. "Does Eastern Europe Belong in NATO?" *Orbis* (Winter 1993): 21–35.

_____. *NATO Enlargement and Central Europe: A Study in Civil Military Relations*. Washington, D.C.: Institute for National Strategic Studies, National Defense University, 1996.

Simon, Jeffrey, ed. *NATO Enlargement: Opinions and Options*. Washington, D.C.: Institute for National Strategic Studies, National Defense University, 1995.

Sloan, Stanley R. *NATO: What Is It?* Congressional Research Service (CRS), Report for Congress no. 97-708F. Washington, D.C.: CRS, Library Of Congress, July 17, 1997.

Solomon, Gerald B. "Prizes and Pitfalls of NATO Enlargement." *Orbis* (Spring 1997): 209–221.

Talas, Peter, Jason Dury, and Sebestyen Gorka. *The Partnership for Peace: The First Year*. Defence Studies no. 8. Budapest: Institute for Strategic and Defence Studies, 1995.

Talbott, Strobe. "Why NATO Should Grow." *The New York Review of Books*, August 10, 1995.

The White House. *Questions and Answers on NATO Enlargement*. Report to Senator Jesse Helms, September 10, 1997.

Index

Albania: reaction to PFP, 35

Albright, Madeleine, 112, 116, 123, 133

Andreatta, Beniamino, 104

Andreotti, Giulio, 8

Androsov, Andrei, 55

Arms control associations: NATO enlargement and, 104–106

Aspin, Les, 28; plan for PFP, 33–35

Associate membership in NATO: proposal for, 20

Atlantic Partnership Council, 107

Baker, James: and NACC, 13, 14; on NATO enlargement, 166; and non-differentiation policy, 9

Balladur, Edouard, 70

Baltic Sea: U.S.-sponsored exercise in, 18

Baltic states: NATO membership for, question of, 109–111, 132; PFP programs in, financing of, 64; relationship with NATO, 95; Russian position on NATO expansion in, 109

Barroso, José Manuel Durao, 80

Bartholomew, Reginald, 16

Bartoszewski, Wladyslaw, 77

Belarus: relationship with NATO, 131

Berger, Sandy, 137

Biden, Joseph, 124

Bingaman, Jeff, 125

Bosnian conflict: effect on Hungary, 45; peacekeeping cooperation in, 17; Russian response to NATO air strikes, 55

Britain. See United Kingdom

Brown, Hank, 50, 65, 66

Brussels summit (1994), 37

Brzezinski, Zbigniew: on NATO membership criteria, 68–69; on NATO-Russia relations, 131; on PFP, 48; on U.S. position on enlargement, 63; on Warsaw Pact, 8, 9

Bulgaria: addition to NATO admission list, 104; application for NATO membership, 45; participation in PFP, 77; PFP presentation by, 45; reaction to PFP, 35, 36; relationship with NATO, 54; and Warsaw Pact, 8

Bush, George, 15

Canada: and NACC, 60

Carrington, Peter, 22

Charette, Hervé de, 94

Cheney, Dick, 10

Chernomyrdin, Viktor, 116

China: NATO outreach and, 125

Chirac, Jacques, 129

Christopher, Warren: on NATO enlargement, 29, 96; at Noordwijk meeting, 78; and PFP, 41, 61

Churkin, Vitali, 54
Ciorbea, Victor, 103
CJTF. *See* Combined Joint Task Forces
Claes, Willy, 21, 85
Clinton, Bill: on Baltic states—NATO relations, 109; on benefits of NATO enlargement, 135–136; on NATO enlargement, 24, 38, 63–64, 72–73, 100; on NATO-Russia Act, 120; Prague speech of, 47, 68; visit to Poland, 62, 67; and Yeltsin, 79, 80–81, 87, 114
CNAD. *See* Conference of National Armaments Directors
Cohen, William, 116–117, 136
Combined Joint Task Forces (CJTF), 33
Conference of National Armaments Directors (CNAD), 40
Conference on Security and Cooperation in Europe (CSCE): "Charter of Paris for a New Europe," 11–12; and German reunification, 6–7; Russian position on, 53
Constantinescu, Emil, 103
Consultative relationship: Eastern European interest in, 54; PFP provisions for, 40–41
Cook, Robin, 104
Copenhagen NAC meeting (1991), 11–12
Corterier, Peter, 9, 50
Costs of NATO enlargement, 98, 101, 121–123, 174
Cragg, Anthony, 85
Crisis management: PFP and, 38
CSCE. *See* Conference on Security and Cooperation in Europe
Czech and Slovak Federal Republic: interest in NATO, 10
Czech Republic: cooperation with NATO, 17; invitation to join NATO. *See* Enlargement of NATO; PFP presentation by, 45; reaction to PFP, 35–36

D'Amato, Alphonse, 103
Daskalov, Stanislav, 36
Davis, Lynn, 28

Defense planning questionnaire, NATO, 89
Dellums, Ronald V., 134
Denmark: on Baltic states-NATO relations, 110
Differentiation policy: opposition to, 20–21; Polish insistence on, 77; of WEU, 14; *See also* Nondifferentiation policy

Eagleburger, Lawrence, 15, 19
EAPC. *See* Euro-Atlantic Partnership Council
Enhanced partnership: areas for, 106–108; and NATO membership, 133
Enlargement of NATO: allies' positions on, 30–32; arguments supporting, 126; benefits of, 123, 135–136, 162; costs of, 98, 101, 121–123, 174; countries considered for, 102–104, 109–111; examination of, 70–73, 85–87; first wave of, 133–134, 136–138; geopolitics and, 47, 68, 133; hurdles to, 94–97; impetus for, 140; NACC as first stage of, 19; negative view of, 124–125; objectives of, 74; PFP language on, 38–40; previous, 2; principles of, 86, 160; priorities in, 141–142; procedures for, 86; rationale for, 86; role of, 134–135; Russian position on, 23–24, 71–72, 79, 112–113, 172; Shea's two-part protocol for, 22; slowing down of, calls for, 80–84; U.S. position on, 30–32, 63–70, 99–100, 123–125, 134–136
ESDI. *See* European Security and Defense Identity
Estonia. *See* Baltic states
EU. *See* European Union
Euro-Atlantic Partnership Council (EAPC), 107; goals of, 108; inauguration of, 132; reactions to, 108–109
European Security Act (1997), 110, 117, 119
European Security and Defense Identity (ESDI), 127
European Union (EU): membership in, and NATO membership, 21

Examination of NATO enlargement, 70–73, 85–86; Polish response to, 87; *See also* Study on NATO Enlargement

Financing: of NATO enlargement, 98, 101, 121–123; of PFP, 42, 63, 64
Finland: PFP presentation by, 45–46
Fischmann, Brent, 170
Folmer, Jan, 85
France: on EAPC, 108; position on NATO enlargement, 32, 70, 94
Freeman, Charles, 27

Genscher, Hans-Dietrich, 13
Geopolitics: and NATO enlargement, 47, 68, 133
Germany: on Baltic states-NATO relations, 110; position on NATO enlargement, 30–31, 32; reunification of, and question of NATO membership, 6–7
Gillmore, David, 31
Gilman, Benjamin, 48, 99
Glenn, John, 125
Gorbachev, Mikhail: coup against, NATO's response to, 12–13; on German reunification, 6, 93; on preservation of Warsaw Pact, 8
Goss, Porter, 126
Goulden, John, 100
Grachev, Pavel, 54
Greece: Bush-Yeltsin meeting of 1992 and, 15
Gromov, Feliks, 172

Hamzik, Pavol, 104
Havel, Vaclav, 7, 26; reaction to PFP, 35–36
Helms, Jesse, 135; on NATO enlargement, 177–178
Helsinki summit, Russia-U.S., 114, 126; misunderstandings arising from, 117; reactions to, 114–116
Holbrooke, Richard, 67, 68, 98
Hungary: and NACC, 60; participation in PFP, 77; PFP presentation by, 45; relationship with NATO, 54,

124–125. *See also* Enlargement of NATO
Hunter, Robert, 39
Hyde, Henry, 48–49

Implementation Force (IFOR), 90
Individual Partnership Program (IPP), 43
Intensified dialogue, 88
IPP. *See* Individual Partnership Program
Italy: position on NATO enlargement, 32, 80
Ivanov, Igor, 132

Jannuzzi, Giovanni, 33
Johnson, Daryl, 28
Joint action: Baker-Genscher proposal for, 13, 26
Juhl, Clarence H., 28
Just, Annette, 126

Karaganov, Sergei, 53, 121
Karkoszka, Andrzej, 18, 96, 99
Kashlev, Yuri, 65
Kempthorne, Dirk, 125
Kennan, George F., 124
Kennedy, Ted, 125
Kinkel, Klaus, 31, 98, 133
Kirchberg Declaration, 50–51
Kissinger, Henry: on Helsinki meeting, 115; on PFP, 47–48; on Warsaw Pact, 8
Knokke-Heist conference (1995), 84
Kobieracki, Adam, 105
Kohl, Helmut; 92
Kostikov, Vyacheslav, 54
Kovac, Michal, 104
Kozyrev, Andrei: on NATO air strikes, 55; on NATO enlargement, 71–72, 80, 81, 85; on NATO-Russia relationship, 90–91; and PFP, 47, 60–61; sacking of, 91, 93
Krenz, Egon, 8
Kruzel, Joseph, 28, 51, 63, 66, 68, 97
Kvitsinsky, Yuli, 11
Kwasniewski, Aleksander, 89

Lake, Anthony, 29
Lamentowicz, Wojciech, 26
Larrabee, F. Stephen, 178
Latvia. See Baltic states
Lebed, Aleksandr, 80, 96–97, 98, 115;
  on NATO enlargement, 126; on
  NATO-Russia Act, 130
Lithuania: demands for membership
  in NATO, 95; See also Baltic states
Litvinov, Maxim, 92
London summit (1990), 9–10
Lugar, Richard, 21, 28, 118, 134; on
  NATO enlargement, 177–178; on
  PFP, 29, 49
Lukin, Vladimir, 55

McCain, John, 50
McCarthy, James P., 27
McCurry, Mike, 136
McDougall, Barbara, 10, 12–13
Madrid summit, 1, 140; clarification of
  issues prior to, 133–134; decision on
  first wave of enlargement, 136–138;
  NATO-Russia Act and, 131; text of
  declaration, 143–147
Mamedov, Georgi, 79
Manilov, Valeri, 56
Matlock, Jack, 166
Meciar, Vladimir, 103
Melanescu, Teodor, 78
Membership in NATO: applicants for
  (by 1996), 89; associate, proposal
  for, 20; "backdoor" approach to, 21;
  benefits of, 2; criteria for, 68–69, 76;
  former Warsaw Pact members' view
  of, 1–2; German reunification and
  question of, 6–7; "Royal Road" approach to, 21
Migranyan, Andranik, 53
Milewski, Jerzy, 46
Military cooperation: PFP and, 40,
  127; Polish aspirations for, 44; pre-
  accession, 163
Mitterrand, François, 8

NAA. See North Atlantic Assembly
NAC. See North Atlantic Council
NACC. See North Atlantic Coopera-
  tion Council

NATO: evolution of, 139–140; purpose of, 3, 75; See also Enlargement
  of NATO; Membership in NATO;
  Special relationship with NATO
NATO Enlargement Facilitation Act,
  99–100
NATO Expansion Act, 48, 75–76
NATO Participation Act, 50, 65, 66–
  67; 1995 Amendments to, 74–75,
  76–77
NATO Revitalization Act, 48
NATO-Russia relationship: clear definition of, importance of, 56; cross-
  guarantees to Eastern Europe, 91,
  92; difficulties in, 71–72; document
  formalizing, 96, 111–120; five-point
  program for cooperation, 98–99;
  Germany's position on, 31; NATO
  enlargement and, 2–3, 39, 86; Polish
  reaction to, 59–60; reaction of
  NATO aspirants to, 79; Russian concerns regarding, 54, 56, 84; Russian
  resistance to, 90–91
NATO-Russia Treaty, 117–120; allies'
  position on, 96; debates on, 96–97,
  111–117; future of, recommendations for, 142; reactions to, 129–
  131; signing of, 129–130; test of,
  141
Naumann, Klaus, 108, 122–123
Netherlands: on EAPC, 108; on
  NATO-Russia Council, 115; position on NATO enlargement, 32
Nondifferentiation policy: in response
  to Warsaw Pact collapse, 7–10
Noordwijk meeting (1995), 74, 77–78,
  81
North Atlantic Assembly (NAA): en-
  dorsement of NATO enlargement,
  69–70; future role of, 142; and Hungary, 45; response to PFP, 50
North Atlantic Cooperation Council
  (NACC): decline of, 107–108; drawbacks of, 16–17, 18; expanded role
  of, proposal for, 16–17, 27; formation of, 13–14; motion to improve,
  16–17, 160; and NATO enlargement, 19; peacekeeping operations
  under, 15, 17–18; versus PFP, 34;

promises of, 19; Russian demands regarding, 57; U.S. recommendations to, 78–79
North Atlantic Council (NAC): 1991 Copenhagen meeting, 11–12; 1994 Istanbul meeting, 58–59; 1995 Noordwijk meeting, 74, 77–78, 81; 1996 Berlin meeting, 94; 1997 Sintra meeting, 131–133
North Atlantic Treaty. See Washington Treaty
Norway: on Baltic states-NATO relations, 110
Nowak, Jerzy, 91
Nuclear weapons question: NATO enlargement and, 96, 104; NATO-Russia Act and, 119
Nunn, Sam, 29, 81–82, 111

Olechowski, Andrzej, 36, 46, 59–60, 71
Organization for Security and Cooperation in Europe (OSCE): commitment of, 118; 1996 Vienna discussions, 91; Yeltsin's proposal regarding, 130
OSCE. See Organization for Security and Cooperation in Europe
Outreach: NATO's policy of, 7–12
Owens, William, 90
Oxman, Stephen, 29–30

PARP. See Planning and Review Process
Partnership Coordination Cell (PCC), 37
Partnership for Peace (PFP): advantages of, 34–35; aftermath to, 46–52; Aspin plan for, 33–35; consultation provisions in, 40–41; disappointment with, 29; enhanced cooperation in, 106–108, 133; enlargement language in, 38–40; exercises under, 77, 90; financing of, 42, 63, 64; framework of, 37–42; versus NACC, 34; objectives of, 19–20, 37; origin of term, 28; presentations for, 43–46; rationale for adopting, 28; reactions of beneficiaries, 35–36;

road to adoption of, 26–30; and Russian military reform, 127; Russian participation in, 56–58, 60–61, 66; Russian reaction to, 35, 47, 53–56; signatories to, 161; strengthening of, 88, 94; U.S. approach to offering, 41–42
Pastusiak, Longin, 23, 78
PCC. See Partnership Coordination Cell
Peace Pannon (IFF program), 77
Peacekeeping: under NACC, 15, 17–18
Perry, William: on NATO enlargement, 92, 160; on PFP, 41, 42
Peters, Hans Jochen, 78
Petersen, Jan, 82–83, 110
Petersen, Niels Helweg, 25
PFP. See Partnership for Peace
Planning and Review Process (PARP): expansion of, 107
PMSC. See Politico-Military Steering Committee
Poland: cooperation with NATO, 17, 44; on differentiation problem, 77; financial support sought by, 63; and NATO membership, 62–63, 89, 95. See also Enlargement of NATO; on NATO-Russia relationship, 59–60; PFP presentation by, 43–45; progress in integration with NATO, 77–78; reaction to PFP, 36, 46–47; response to NATO enlargement study, 87; Russian response to NATO-membership aspirations of, 23–24; and Warsaw Pact, 8
Politico-Military Steering Committee (PMSC), 43; Russian request for participation in, 57
Portillo, Michael, 92
Portugal: Bush-Yeltsin meeting of 1992 and, 15; position on NATO enlargement, 80
Primakov, Yevgeni, 93, 97, 98, 112
Pszczel, Robert, 60, 77, 95
Public opinion: as hurdle to NATO enlargement, 97; of NATO enlargement, in Russia, 126, 176, 177

Republican Congress: and NATO expansion, 67

Rifkind, Malcolm, 30, 96
Rodionov, Igor, 97, 127
Rodman, Peter, 87
Rogov, Sergei, 56, 84–85
Romania: bilateral partnership proposed by, 26; PFP presentation by, 45; prospects for NATO membership, 103, 136, 137; reaction to PFP, 35
Rosati, Dariusz, 95, 130
Rosner, Jeremy, 127–128
Roth, William V., Jr., 69, 102–103, 137
Rühe, Volker: on Baltic states–NATO relations, 110; on NATO enlargement, 21, 30, 66; on NATO-Russia relationship, 31; on PFP, 43
Russia: armed forces of, chaos in, 127; concessions to, 85; cooperation with NATO in Bosnia, 17; domestic policies of, NATO enlargement and, 84, 92–94; first NATO visit to, 9; Madrid summit and, 138; membership in NATO, question of, 13, 66, 111; military doctrine of, draft, 14–15; misunderstanding about NATO, 116; NAC approach to, 59; and NATO. *See* NATO-Russia relationship; NATO aspirants and, 65; obstructionism of, 79–80; participation in PFP, 56–58, 60–61, 66; perspectives on NATO's role, 97–98; position on Baltic states–NATO relations, 109; position on NATO enlargement, 23–24, 71–72, 79, 112–113, 172; 1996 presidential elections in, run-up to, 91, 92; public opinion of NATO enlargement, 126, 176, 177; reaction to PFP, 35, 47, 53–56; reassurance to, 104; special treatment of, 71; *Study on NATO Enlargement* and, 85–86; and Ukraine, friendship treaty between, 131; and WEU, 51
Rybkin, Ivan, 97

Schaefer, Helmut, 133
Scowcroft, Brent, 115

Security: as goal of NATO enlargement, 74
Serov, Valeri, 138
Shea, Jamie, 77; protocol for NATO enlargement, 22
Sheehan, John J., 78
Shevardnadze, Eduard: on German reunification, 6–7; on NATO enlargement, 166
Shustov, Vladimir, 91
Simon, Jeffrey, 19–20, 63
Simon, Paul, 50, 65
Sintra meeting of NAC (1997), 131–133
Slovakia: consideration for NATO membership, 103–104; *See also* Czech and Slovak Federal Republic
Slovenia: prospects for NATO membership, 136, 137
Smith, Christopher H., 103
Smith, William, 39
Solana, Javier, 112, 121, 130, 131, 132, 137
Slocombe, Walter, 42
Soviet Union: perceived assurances of NATO nonenlargement, 165–166; strategy to preserve status quo, 11
Spain: Bush-Yeltsin meeting of 1992 and, 15; rescue exercises of 1993 and, 17–18
Special relationship with NATO: Polish demands for, 62–63; Russia and, 71; Ukraine and, 120–121, 131
Spence, Floyd D., 134
Study on NATO Enlargement, 91, 104; Polish response to, 87; Russia and, 85–86; *See also* Examination of NATO enlargement
Surkov, Mikhail, 116
Sweden: on EAPC, 109; and PFP, 46

Taft, William, 19
Talbott, Strobe, 28, 33, 79, 93
Thatcher, Margaret, 8
Tinca, Gheorghe, 45
Torricelli, Robert G., 171
Towpik, Andrzej, 44
Travemünde meeting (1993), 33

Turkey: Bush-Yeltsin meeting of 1992 and, 15

Ukraine: cooperation with NATO, 17; NAC approach to, 59; position on NATO enlargement, 95–96; and Russia, friendship treaty between, 131; special relationship with NATO, 120–121, 131
United Kingdom: position on NATO enlargement, 30, 32
United States: leisurely approach to NATO enlargement, 87; and PFP, 34, 41–42, 47–50; position on NATO enlargement, 30–32, 63–70, 99–100, 123–125, 134–136; public opinion of NATO enlargement, 126; recommendations to NACC, 78–79; role in NATO evolution, 140–141; support for NATO enlargement, 99–100, 123; WEU expansion and, 51–52

van Eekelen, Willem, 51, 86
van Traa, Maarten, 115
Vershbow, Alexander, 63, 67, 92
Visegrad meeting, 7
Voevoda, Yuri, 56, 80
von Moltke, Gebhardt, 56, 85, 89, 94–95, 105–106
von Richthofen, Hermann, 70

Walesa, Lech, 23; on PFP, 46
Warner, John, 125
Warsaw Pact: attempts to preserve, 8–9; basis of, 6; collapse of, 7; NATO's response to collapse of, 7–12; Romania in, 103

Washington Treaty, 75, 130; Article 5 of, 40; Article 10 of, 13, 20; preamble to, 2
Wegener, Henning, 8
Western European Union (WEU), 10; as alternative to NATO expansion, 125–126; differentiated approach of, 14; enlargement of, 50–52, 125–126
WEU. See Western European Union
Wisniewski, Grzegorz, 166
Wörner, Manfred: at 1992 NACC meeting, 17; on NATO enlargement, 18, 24–25; on NATO transformation, 139; and nondifferentiation policy, 7–8; visit to Moscow, 9
Wozniak, Stanislaw, 44

Yeltsin, Boris: on Baltic states–NATO relations, 109; and Clinton, 79, 80–81, 87, 114; on NATO enlargement, 71–72, 93, 112–113, 129; on NATO membership for Russia, 13; on OSCE, 130; on PFP, 35; on Polish membership in NATO, 23–24; in run-up to 1996 elections, 91; and U.S.-Russian relations, 15
Yugoslavia: conflict in. See Bosnian conflict
Yuriev, Mikhail, 125–126
Yushenkov, Sergei, 55, 65

Zatulin, Konstantin, 55
Zhelev, Zhelyu, 45
Zieleniec, Jozef, 36
Zoellick, Robert, 47, 93–94
Zyuganov, Gennadi, 114, 120

About the Author

Representative Gerald B. Solomon (R-N.Y.) is chairman of the Rules Committee of the U.S. House of Representatives and member of the House Task Force on National Defense Policy and the Task Force on American Prisoners of War and Missing in Southeast Asia. He is vice president of the North Atlantic Assembly, the NATO parliamentarians' organization. He is a ten-term member of the U.S. Congress from the Hudson Valley and Adirondack region of upstate New York. In September 1997 the Speaker of the House of Representatives appointed Congressman Solomon as chairman of the House NATO Observers' Group, which plays a vital role in securing congressional support for NATO and its enlargement.

Congressman Solomon previously served as senior ranking member of the House Veterans Committee and as a veteran member of the House Foreign Affairs Committee. He was designated by President Ronald Reagan as ambassador designate to the United Nations and congressional adviser to the United Nations Special Session on Disarmament.

After enlisting in the U.S. Marine Corps at the onset of the Korean War, Congressman Solomon served on active duty until July 25, 1952, and remained in the U.S. Marine Corps reserve until honorably discharged on July 14, 1959.

This volume was prepared as part of the author's work as a government employee.